FM 6-81

DEPARTMENT OF THE ARMY FIELD MANUAL

155-mm
HOWITZER M1
FIELD MANUAL

SERVICE OF THE PIECE

By **DEPARTMENT OF THE ARMY** • APRIL 1948

DISCLAIMER:

This document is a reproduction of a text first published by the Department of the Army, Washington DC. All source material contained herein has been approved for public release and unlimited distribution by an agency of the US Government. Any US Government markings in this reproduction that indicate limited distribution or classified material have been superseded by downgrading instructions promulgated by an agency of the US government after the original publication of the document No US government agency is associated with the publication of this reproduction. This manual is sold for historic research purposes only, as an entertainment. It contains obsolete information and is not intended to be used as part of an actual training program. No book can substitute for proper training by an authorized instructor.

DEPARTMENT OF THE ARMY FIELD MANUAL
FM 6-81

This manual supersedes FM 6-81, 23 December 1943, including C 1, 28 October 1944, and C 2, 30 May 1945.

SERVICE OF THE PIECE

155-mm
HOWITZER M1

TRUCK-DRAWN
AND
TRACTOR-DRAWN

DEPARTMENT OF THE ARMY • *APRIL 1948*

United States Government Printing Office
Washington : 1948

DEPARTMENT OF THE ARMY
Washington 25, D. C., 21 April 1948

FM 6–81, Service of the Piece, 155–mm Howitzer M1, Truck-Drawn and Tractor-Drawn, is published for the information and guidance of all concerned.

[AG 300.7 (22 Jan 48)]

By order of the Secretary of the Army:

OMAR N. BRADLEY
Chief of Staff, United States Army

Official:
EDWARD F. WITSELL
Major General
The Adjutant General

Distribution:
Army:
Arm & Sv Bd (1); AGF (40); MDW (3); A (ZI) (17), (Overseas) (2); CHQ (2); D (5); B 6 (2); R 6 (2); USMA (5); Sch (2) except Sch 6 (500); Tng Ctr (2); PG 9 (2); T/O & E 6–36 (5); 6–37 (10); 6–39 (5); 6–337 (10); 6–339 (5); SPECIAL DISTRIBUTION.
Air Force:
USAF (2).
For explanation of distribution formula see TM 38–405.

CONTENTS

This manual supersedes FM 6–81, 23 December 1943, including C 1, 28 October 1944, and C 2, 30 May 1945.

CHAPTER 1

GENERAL

---

1. PURPOSE AND SCOPE. This manual prescribes the duties to be performed in the service of the piece by personnel normally assigned to one howitzer section of the firing battery.

2. REFERENCES. **a. Description, operation, functioning, and care of matériel.** TM 9–331, ORD 7 SNL C–39, ORD 8 SNL C–39, and ORD 9 SNL C–39.

b. Description and operation of sighting equipment. TM 9–331, TM 9–1582, ORD 7 SNL F–214, ORD 8 SNL F–214, ORD 9 SNL F–214, ORD 7 SNL F–216, ORD 8 SNL F–216, and ORD 9 SNL F–216.

c. Ammunition. TM 9–331, TM 9–1900, TM 9–1901, ORD 11 SNL R–1, ORD 11 SNL R–2, ORD 11 SNL R–6, and FT 155–Q–2.

d. Cleaning and preserving materials. TM 9–850 and ORD 3 SNL K–1.

e. Automotive driver. FM 25–10, TM 21–300, TM 21–301, TM 21–305, and TM 21–306.

f. Safety precautions in firing. AR 750–10, FM 6–140, and TM 9–1900.

g. Firing battery. FM 6–140.

h. Gunnery. FM 6–40.

For military terms not defined in this manual, see TM 20–205; for list of training publications, see FM 21–6; for training films, film strips, and film bulletins, see FM 21–7; for training aids, see FM 21–8.

i. Reconnaissance, occupation, and organization of position. FM 6–20, FM 6–101, and FM 6–140.

j. Defense against chemical attack. FM 21–40.

k. Decontamination of matériel. TM 3–220.

3. DEFINITIONS AND TERMS. **a. Section.** Tables of organization and equipment prescribe the personnel and matériel comprising a section of a battery. In this manual the term is frequently used to designate the personnel required to serve one piece and the matériel of the section. In a restricted sense, the term section may be used to designate *only the personnel* of the section.

b. Coupled. A piece is coupled when its lunette is attached to the pintle of its prime mover.

c. Uncoupled. A piece is uncoupled when its lunette is detached from the pintle of its prime mover.

d. Front. The front of a section, piece coupled is the direction in which the prime mover is headed; piece uncoupled, the direction in which the muzzle points. However, for determining the right or left of the piece, coupled or uncoupled, the front is the direction in which the muzzle points.

e. Right (left). The direction right (left) is the right (left) of one facing to the front.

f. In battery. A howitzer is said to be in battery when its tube is in normal firing position.

CHAPTER 2

ORGANIZATION

---

4. COMPOSITION OF HOWITZER SECTION. The personnel of a howitzer section consists of—

    a. A chief of section.
    b. A gunner.
    c. Eight cannoneers, numbered from 1 to 8.
    d. An ammunition corporal.
    e. A driver.

5. FORMATION OF HOWITZER SECTION. a. Order of formation. The personnel of a howitzer section are formed as in figure 1. Higher-numbered cannoneers, if present, form in order between No. 8 and the ammunition corporal. The driver, if present, forms to the left of the ammunition corporal.

    b. To form. (1) The place of formation is indicated and the commands given thus, for example: 1. IN FRONT (REAR) OF YOUR PIECES, or 1. ON THE ROAD FACING THE PARK, 2. FALL IN. The gunner repeats the command **FALL IN** and hastens to place himself, faced in the proper direction, at the point where the right of the section is to rest. The cannoneers move in double time and assemble at attention in their proper places. For the first formation of the howitzer sections for any drill or exercise, the caution, "As howitzer sections," precedes the command. The chief of section, if present, supervises the formation. His post is two paces in front of the center of the section.

Figure 1. Formation of howitzer section.

(2) In case the front or rear of the piece is designated, each section falls in at its post (par. 6).

c. To call off. (1) The command is: CALL OFF. The cannoneer on the left of the gunner calls off "One"; the cannoneer on the left of No. 1, "Two"; and so on. The gunner, the ammunition corporal, and the driver do not call off.

(2) After having called off, if a subsequent formation is ordered, the members of the section fall in at once in their proper order.

CHAPTER 3

POSTS: MOUNTING AND DISMOUNTING

6. POSTS OF HOWITZER SECTION. **a. Piece coupled.** (1) *In front of piece.* The section is in line facing to the front, its rear and center two paces from the front of the prime mover (fig. 2).

Figure 2. Post of section, piece coupled, in front of piece.

(2) *In rear of piece.* The section is in line facing to the front, the chief of section one pace from the muzzle of the piece (fig. 3).

Figure 3. Post of section, piece coupled, in rear of piece.

b. Piece uncoupled, in rear of piece. (1) *Piece not prepared for action.* The section is in line facing to the front, the chief of section one pace from the ends of the trails (fig. 4).

Figure 4. Post of section, piece uncoupled, in rear of piece, piece not prepared for action.

(2) *Piece prepared for action.* The section is in line facing to the front, the chief of section one pace to the rear of the midpoint between the ends of the trails (fig. 5).

c. Piece uncoupled, in front of piece. With the piece either prepared for action or not prepared

Figure 5. Post of section, in rear of piece, piece prepared for action.

9

Figure 6. Post of section, piece uncoupled, in front of piece.

for action, the section is in line facing to the front, its rear and center two paces from the muzzle of the piece (fig. 6).

7. TO POST HOWITZER SECTIONS. The sections, having been marched to the vicinity of the pieces, are posted at the command SECTIONS IN FRONT (REAR) OF YOUR PIECES, FALL IN. Each chief of section marches his section to its piece and posts it in the position indicated.

8. POSTS OF CANNONEERS. a. Piece coupled. The members of the howitzer section are posted as in

Figure 7. Posts of cannoneers, piece coupled.

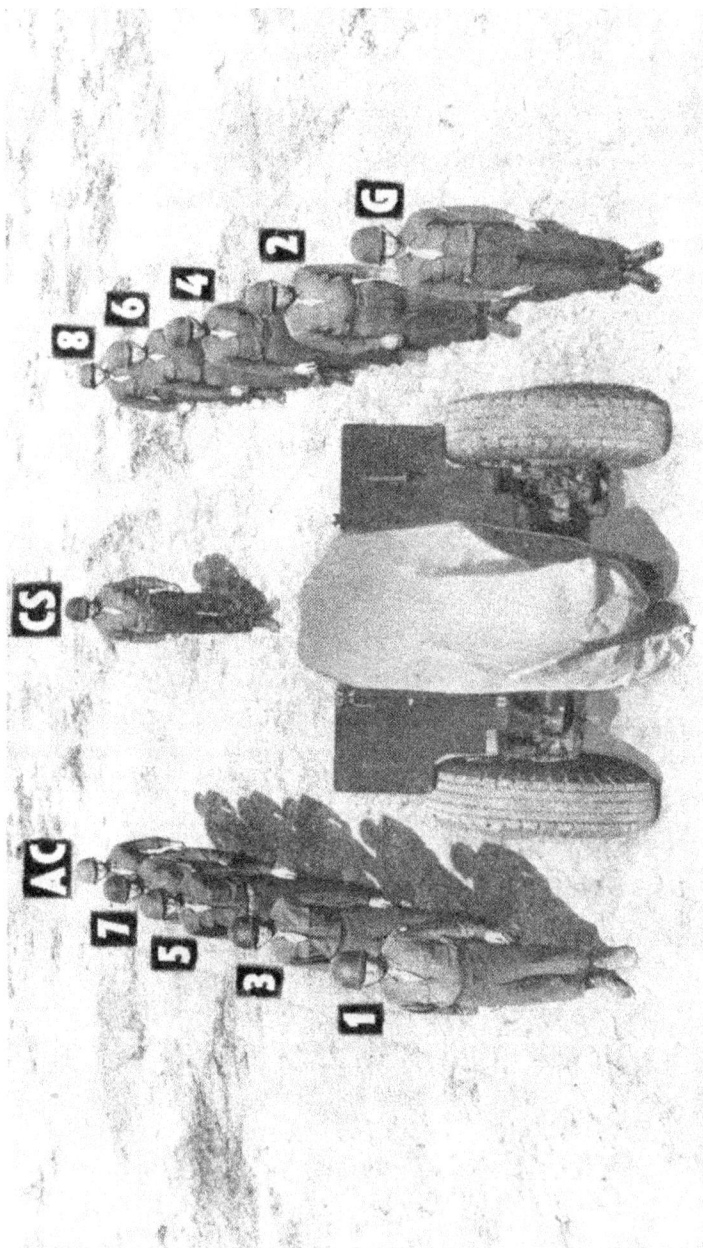

Figure 8. Posts of cannoneers, piece uncoupled, not prepared for action.

figure 7. All are 2 feet outside the wheels and facing to the front.

b. Piece uncoupled, not prepared for action. The posts of the members of the howitzer section are shown in figure 8. All are facing to the front.

c. Piece prepared for action. The piece having been uncoupled and prepared for action, the posts of the members of the section are as follows:

(1) *Chief of section.* The chief of section goes where he can control the service of the piece, hear commands, and perform his duties effectively. A convenient post is near the trail spade opposite the executive.

(2) *Gunner, ammunition corporal, and cannoneers.* The posts are shown in figure 9. Higher-numbered cannoneers, if present, take posts as prescribed by the chief of section.

9. TO POST CANNONEERS. **a.** The commands are: 1. CANNONEERS, 2. POSTS. Each gunner repeats the command POSTS. The cannoneers move at double time to their posts.

b. For preliminary instruction, the sections on entering the park are first posted with their pieces, and the cannoneers are then sent to their posts by the foregoing command. The command is general, however, and applies when cannoneers are in or out of ranks, at a halt or marching, and when the pieces are coupled or uncoupled.

c. At drill, all stand at attention at their posts facing the front. In firing and combat, minor modifications of these posts are required for the more efficient performance of duties in service of the piece, and for protection of personnel. Higher-numbered

Figure 9. Posts of cannoneers, piece in firing position, prepared for action.

cannoneers, if present, take posts as prescribed by the chief of section.

d. In order to exercise the cannoneers in all duties connected with the service of the piece and lend variety to the drill, posts of individual cannoneers should be changed frequently. The cannoneers being at their posts, piece coupled or uncoupled, the commands are: 1. CHANGE POSTS, 2. MARCH. In each section No. 1 quickly takes the post of No. 2, No. 2 of No. 3, and so on, No. 8 taking the post of No. 1. The gunner and higher-numbered cannoneers change only when specifically directed. The commands are: 1. GUNNER AND CANNONEERS CHANGE POSTS, 2. MARCH. In each section, the gunner quickly takes the post of No. 1, No. 1 of No. 2, and so on. The highest numbered cannoneer takes the post of the gunner.

10. TO MOUNT CANNONEERS. a. Tractor prime mover. The commands are: 1. CANNONEERS, PREPARE TO MOUNT, 2. MOUNT. At the first command the members of the howitzer section, except the chief of section and driver, take positions as shown in figure 7; the driver opens the doors on the left side of the prime mover and stations himself opposite and 2 feet from the left front door facing to the rear; the chief of section opens the doors on the right side of the prime mover and stations himself opposite and 2 feet from the right front door facing to the rear. At the second command the gunner and Nos. 1 to 7 inclusive, from their respective sides, mount in order from front to rear and take seats as shown in figure 10; No. 8 and the ammunition corporal move at double time to the vehicle

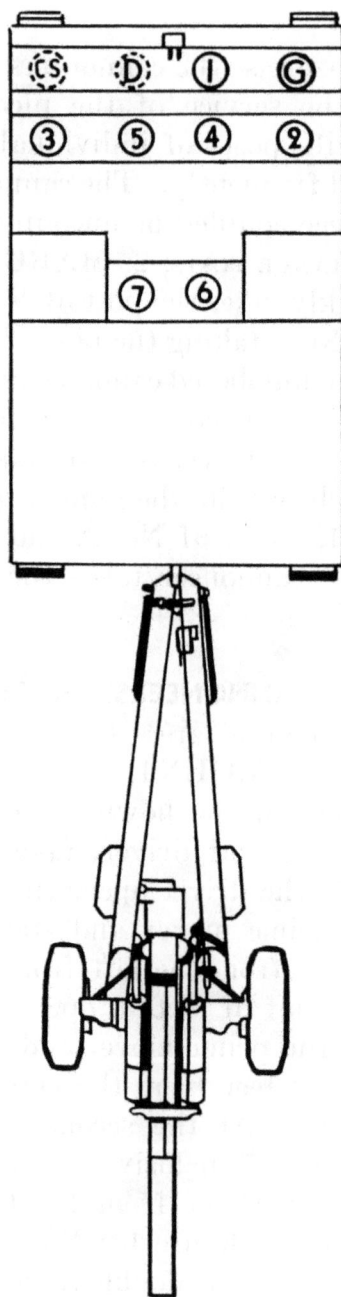

Figure 10. Section mounted, tractor prime mover.

designated to carry them. If the chief of section and driver are to be included in the movement, the commands are: 1. PREPARE TO MOUNT, 2. MOUNT. At the second command, and after the gunner and cannoneers have mounted, the driver and chief of section take their seats as shown in figure 10, both mounting from the left.

b. Truck prime mover. The commands are: 1. CANNONEERS, PREPARE TO MOUNT, 2. MOUNT. At the first command, the members of the section move at double time to positions shown in figure 7. At the second command, cannoneers of both columns mount in order from front to rear, and take seats as in figure 11. Each cannoneer is assisted by the one directly behind (or in front, in the case of the last cannoneer in column) to insure promptness and prevent injuries. If the chief of section and driver are to be included in the movement, the commands are: 1. PREPARE TO MOUNT, 2. MOUNT. At the first command, the driver and chief of section take position on the left and right of the rear of the truck, respectively. After the last cannoneer has mounted, the driver fastens the safety strap; the chief of section and driver mount, take seats, and close their doors.

c. If the commands are: 1. CANNONEERS, 2. MOUNT, the cannoneers execute, at the command MOUNT, all that has been prescribed for the commands CANNONEERS, PREPARE TO MOUNT and

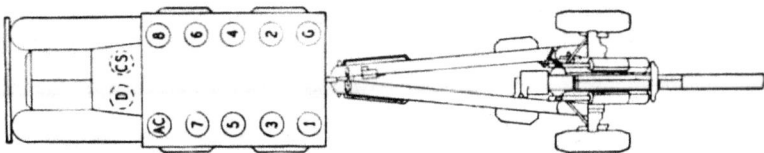

Figure 11. Section mounted, truck prime mover.

MOUNT. If the chief of section and driver are to be included in this movement, the command is: MOUNT.

11. TO DISMOUNT CANNONEERS. a. The commands are: 1. CANNONEERS, PREPARE TO DISMOUNT, 2. DISMOUNT. At the first command, cannoneers assume positions from which they can dismount promptly; at the second command they jump to the ground and take their posts at double time. If the chief of section and driver are to be included in this movement, the commands are: 1. PREPARE TO DISMOUNT, 2. DISMOUNT.

b. If the commands are: 1. CANNONEERS, 2 DISMOUNT, the cannoneers execute at the command DISMOUNT all that has been prescribed for the commands CANNONEERS, PREPARE TO DISMOUNT and DISMOUNT. If the chief of section and driver are to be included in this movement, the command is: DISMOUNT.

CHAPTER 4

TO PLACE HOWITZER IN FIRING AND TRAVELING POSITIONS

---

12. MOVEMENT OF PIECE BY HAND. The weight of the 155–mm howitzer restricts movements by hand by the howitzer section beyond those required in uncoupling, preparing for action, resuming march order, and coupling. When additional movement by hand is necessary it will be done with additional personnel provided by, and under the supervision of the executive.

13. TO UNCOUPLE. a. General. At drills, prime movers are posted as directed by the battery commander. In combat and in instruction and drill simulating combat, the first sergeant conducts the prime movers to a place previously designated by the battery commander. There they are disposed so as to take best advantage of cover and concealment; if neither is available, the prime movers are located in rear of either flank with wide intervals between them.

b. To fire to the front. The command is: ACTION FRONT. If marching, the prime mover halts at the command or signal. The section (less driver), if mounted, dismounts after the prime mover has halted.

(1) *Piece.* The gunner and No. 1 hasten to the wheels of the piece and set the left- and right-hand brakes respectively. (See par. 3d.) Nos. 2 to 8 inclusive hasten to the trail handles, preparatory to un-

coupling. Nos. 7 and 8 close the service and emergency cut-out cocks respectively on the prime mover. The gunner then opens the drain cock on the emergency air tank of the piece. Nos. 7 and 8 pass the air hoses back to Nos. 6 and 5 respectively, who couple the hoses to the dummy couplings on the trails of the piece. No. 7 unlatches the pintle and, assisted by Nos. 2, 3, 4, 5, 6, and 8, raises the lunette from the pintle of the prime mover. The gunner and No. 1 assist in the lifting by placing their weight on the muzzle end of the tube. At the chief of section's command or signal, the driver moves the prime mover forward a few feet. No. 1 releases the right-hand brake. Nos. 2 to 8 inclusive then swing the piece clockwise 180° and lower the trails to the ground. No. 1 sets the right-hand brake. Assisted by the ammunition corporal and the gunner, all cannoneers then unload the ammunition, tools, and accessories and arrange them in an orderly and convenient manner. When the unloading has been completed, the chief of section commands or signals DRIVE ON. The piece is immediately prepared for action as prescribed in paragraph 14.

(2) *Prime mover.* At the command DRIVE ON, the prime mover is moved out and posted as prescribed in a above.

c. To fire to the rear. The command is: ACTION REAR. This is executed in the same manner as for ACTION FRONT except that the piece is not turned except for small shifts as may be directed by the chief of section.

d. To fire to the flank. The command is: ACTION RIGHT (LEFT). The movement is executed according to the principles of ACTION FRONT, with the following modification: Nos. 2 to

8 inclusive swing the piece clockwise 90° for AC-TION LEFT or counterclockwise 90° for ACTION RIGHT, or until the muzzle is pointed in the direction of fire prescribed by the chief of section.

e. To fire from the march. The prime movers halt at the command or signal ACTION FRONT (REAR) (RIGHT) (LEFT). The movements described in **b, c,** or **d** above are executed without additional commands.

14. TO PREPARE FOR ACTION. See figures 12 to 15, inclusive, and TM 9–331. **a. General.** The piece being in position uncoupled, the command is: PREPARE FOR ACTION.

b. Duties of individuals of the howitzer section. (1) *Chief of section.* (*a*) Supervises the work of the cannoneers.

(*b*) Places himself between trails facing to the rear, grasps jack float, and commands SPREAD.

(*c*) Places jack float outside left trail.

(*d*) Locks right spade in firing position with spade key.

(*e*) Verifies the direction of fire before trail pits are dug and commands MARK TRAILS.

(*f*) Inspects the matériel, determines that all is in order, and when the operations have been completed, reports to the executive, "Sir, No. (so-and-so) in order," or reports any defects which the section cannot remedy promptly.

(2) *Gunner.* (*a*) Locks left spade in firing position with spade key.

(*b*) Assisted by Nos. 1 and 2, removes rear section of howitzer cover.

(*c*) Lowers top portion of left shield when necessary.

Figure 12. Preparation for action prior to spreading trails.

(*d*) Assists No. 5 to unlock traveling lock by manipulating elevating handwheel, that elevates tube to approximately 400 mils.

(*e*) Removes telescope from telescope case, places it in the telescope mount, and sets all scales of panoramic telescope at zero.

(*f*) Uncovers telescope mount bubbles, sets elevation at 400 mils, and levels the bubbles.

(*g*) Closes drain cock to air tank.

(*h*) Takes his post.

(3) *No. 1.* (*a*) Unlatches trial handspike support spring lever on right trail.

(*b*) Assisted by No. 3, removes spade from traveling position on right trail and places it in the approximate position of end of right trail when spread.

(*c*) Assists Nos. 3, 5, and 7 to spread right trail.

(*d*) Moves to left of spade and, assisted by No. 3,

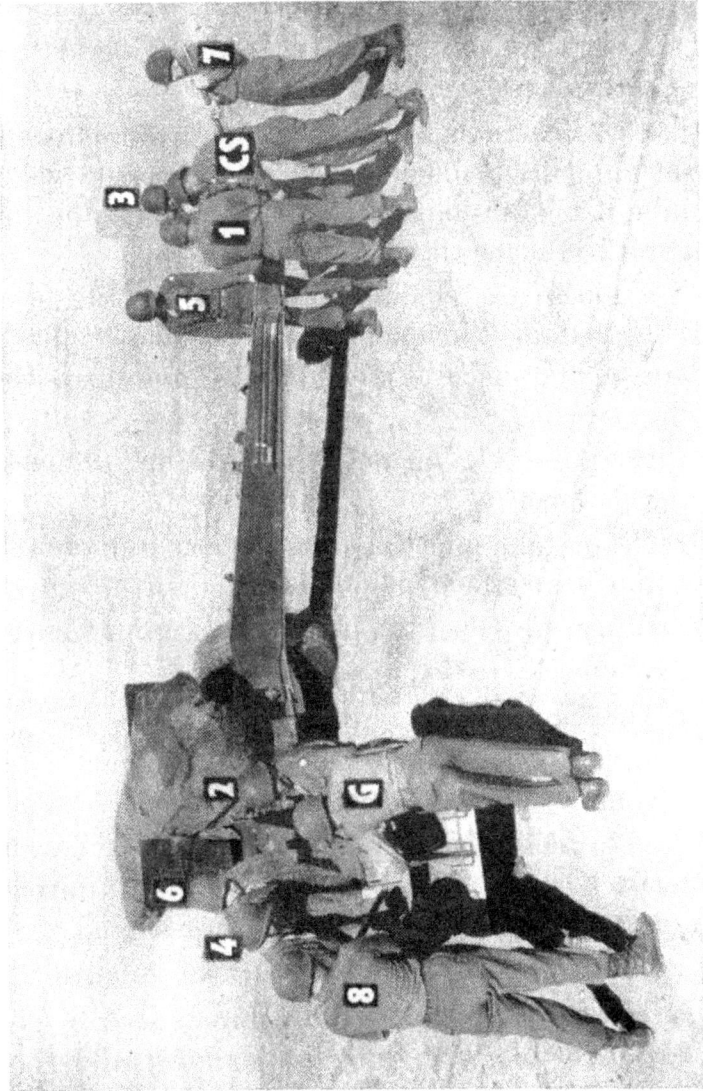

Figure 13. Preparation for action after trails are spread.

23

raises spade into firing position on right trail before trail is lowered.

(*e*) After spade pit is marked by No. 7, assists Nos. 3, 5, and 7 to partially close right trail.

(*f*) Assists gunner and No. 2 to remove rear section of howitzer cover.

(*g*) Procures the lanyard, spare firing mechanism, vent-cleaning bit, primer-seat cleaning reamer, waste, and oiler from section chest and places them on the right trail six inches in rear of the axle.

(*h*) Unlocks the percussion hammer locking pin, locks the hammer in open position, removes firing mechanism and places it on right trail, and opens the breech.

(*i*) Examines the primer vent, the breechblock, and the chamber.

(*j*) Cleans and oils the breechblock and breech recess and cleans the primer vent.

(*k*) When directed by chief of section, assists other cannoneers to seat spades in pits.

(*l*) Takes his post.

(4) *No. 2.* (*a*) Unlatches trail handspike support spring level on left trail.

(*b*) Assisted by No. 4, removes spade from traveling position on left trail and places it in the approximate position of end of left trail when spread.

(*c*) Assists Nos. 4, 6, and 8 to spread left trail.

(*d*) Moves to right of spade and, assisted by No. 4, raises spade into firing position on left trail before trail is lowered.

(*e*) After spade pit is marked by No. 8, assists Nos. 4, 6, and 8 to partially close trail.

24

(*f*) Assists gunner and No. 1 to remove rear section of howitzer cover, folds it, and places it three feet to the right of the right wheel.

(*g*) Inspects spade keys and seats them with sledge if necessary.

(*h*) Procures swabbing sponge, burlap, and pail of water, and places them just forward of the spade along inner side of left trail.

(*i*) Procures loading-rammer head from section chest, assembles it to two sections of the rammer staff, and places it beside the trail to the left of his post.

(*j*) Cleans powder chamber and, assisted by No. 4, cleans bore if necessary.

(*k*) When directed by chief of section, assists other cannoneers to seat spades in pits.

(*l*) Takes his post.

(5) *No. 3.* (*a*) Assists No. 1 to remove the spade from traveling position on right trail and place it in approximate position of end of right trail when spread.

(*b*) Assists Nos. 1, 5, and 7 to spread the right trail.

(*c*) Assists No. 1 to raise the spade into firing position on right trail before trail is lowered.

(*d*) After spade pit is marked by No. 7, assists Nos. 1, 5, and 7 to partially close right trail.

(*e*) Obtains the fuze setter and fuze wrench from the section chest and places them conveniently near the ammunition.

(*f*) When directed by chief of section, assists other cannoneers to seat spades in pits.

(*g*) Assists ammunition corporal in preparation of fuzes and ammunition.

(*h*) Takes his post.

(6) *No. 4.* (*a*) Assists No. 2 to remove spade from traveling postion on left trail and place it in the approximate position of end of left trail when spread.

(*b*) Assists Nos. 2, 6, and 8 to spread left trail.

(*c*) Assists No. 2 to raise the spade into firing position on left trail before trail is lowered.

(*d*) After spade pit is marked by No. 8, assists other cannoneers to partially close left trail.

(*e*) Takes the jack float and places it under the jack with the slot to the front.

(*f*) Assists Nos. 5 and 6 to remove front section of howitzer cover.

(*g*) Working with No. 5, jacks piece into firing position (Carriage M1A1).

    1. Removes firing-jack travel-locking lever.

    2. Lowers firing-jack rack plunger far enough for No. 5 to remove firing-jack housing bottom cover.

    3. Continues to lower rack plunger until ball at lower end engages socket on firing-jack float.

    4. Assisted by No. 5, raises float into position on end of plunger and rotates float 90° counterclockwise, which leaves beveled edge of float to the rear.

    5. Assists No. 5 to jack piece up until firing-jack key can be inserted in housing.

    6. Lowers piece until rack plunger contacts the firing-jack key.

(*h*) Working with No. 5, jacks piece into firing position (Carriage M1A2).

    1. Assists No. 5 to release firing jack from traveling position and place it in firing position.

2. Lowers firing-jack plunger until ball at lower end of plunger engages socket on firing-jack float.

3. Assisted by No. 5, raises float into position on end of plunger and rotates float 90° counterclockwise, which leaves beveled ' edge of float to the rear.

4. Assists No. 5 to jack piece up to limit of plunger travel.

5. Locks plunger in firing position.

(*i*) Passes jack handle to No. 5.

(*j*) Assists No. 2 to clean bore if necessary.

(*k*) Takes his post.

Figure 14. Preparation for action, lowering plunger prior to jacking piece into firing position.

Figure 15. Preparation for action, jacking piece into firing position.

(7) *No. 5.* (*a*) Removes right trail handspike and passes it to No. 7.

(*b*) Assists Nos. 1, 3, and 7 to spread right trail.

(*c*) Assists No. 7 to hold right trail up while spade is placed in firing position.

(*d*) After spade pit is marked by No. 7, assists Nos. 1, 3, and 7 to partially close trail.

(*e*) Passes to the right of the piece and assists Nos. 4 and 6 to remove front section of howitzer cover.

(*f*) Removes jack handles and places them on ground in front of the float.

(*g*) Working with No. 4, jacks piece into firing position (Carriage M1A1).

1. Removes traveling-lock pin as gunner manipulates elevating handwheel and lowers traveling lock to firing position.

2. When No. 4 has lowered firing-jack rack plunger, removes firing-jack housing bottom cover.

3. As rack plunger is lowered, alines float so that ball at lower end of plunger engages socket on firing-jack float.

4. Assists No. 4 to raise float into position on end of jack plunger and to rotate float.

5. Assisted by No. 4, jacks piece up until firing-jack key can be inserted in housing.

6. Inserts firing-jack key and directs No. 4 to lower piece until rack plunger contacts firing-jack key.

(*h*) Working with No. 4, jacks piece into firing position (Carriage M1A2).

1. Removes firing-jack locking pin.

2. Assisted by No. 4, releases firing jack from traveling position and secures it in firing position.

3. Removes traveling-lock pin as gunner manipulates elevating handwheel and lowers traveling lock to firing position.

4. As firing-jack plunger is lowered, alines float so that ball at lower end of plunger engages socket on firing-jack float.

5. Assists No. 4 to raise float into position on end of jack plunger and to rotate float.

6. Assisted by No. 4, jacks piece up to limit of plunger travel.

7. Manipulates ratchet so No. 4 can lock plunger in firing position.

(*i*) Receives jack handle from No. 4 and replaces handles on shield.

(*j*) Secures two firing mechanisms from right trail, inspects, cleans, and oils them.

(*k*) Prepares primers for firing.

(*l*) Takes his post.

(8) *No. 6.* (*a*) Removes left trail handspike, loosens spade keys, and passes handspike to No. 8.

(*b*) Removes spade keys from left trail and places one on each trail in front of the trail lock.

(*c*) Assist Nos. 2, 4, and 8 to spread left trail.

(*d*) Assists No. 8 to hold left trail up while spade is placed in firing position.

(*e*) After spade pit is marked by No. 8, assists Nos. 2, 4, and 8 to partially close trail.

(*f*) Assisted by Nos. 4 and 5, removes front section of howitzer cover, folds, and places it 3 feet to the right of the right wheel.

(*g*) Removes muzzle cover and places it on howitzer cover.

(*h*) Assembles four sections of the rammer staff and bore brush and places the bore brush on the howitzer cover to the right of the piece, with staff to the rear.

(*i*) Removes aiming posts from right trail, assembles them, and places them on howitzer cover. Sets out aiming posts when directed by chief of section (par. 41).

(*j*) Assists ammunition corporal in the preparation of powder charges.

(*k*) When directed by chief of section, assists other cannoneers to seat spades in pits.

(*l*) Takes his post.

(9) *No. 7.* (*a*) Unlocks the trail lock.

(*b*) Receives handspike from No. 5 and inserts it in handspike socket on spade end of right trail.

(*c*) From the left of the handspike, assisted by Nos. 1, 3, and 5, spreads right trail.

(*d*) Assisted by No. 5, holds the right trail up while spade is placed in firing position.

(*e*) Marks right spade pit when directed.

(*f*) Assisted by Nos. 1, 3, and 5, partially closes trail.

(*g*) Digs right spade pit.

(*h*) When directed by chief of section, assists other cannoneers to seat spades in pits.

(*i*) Replaces right trail handspike in bracket on right trail.

(*j*) Assists ammunition corporal in the preparation of ammunition.

(*k*) Takes his post.

(10) *No. 8.* (*a*) Removes trail lock pin.

(*b*) Receives handspike from No. 6 and inserts it in handspike socket on spade end of left trail.

(*c*) From the right of the handspike, assisted by Nos. 2, 4, and 6, spreads left trail.

(*d*) Assisted by No. 6, holds the left trail up while spade is placed in firing position.

(*e*) Marks left spade pit when directed.

(*f*) Assisted by Nos. 2, 4, and 6, partially closes trail.

(*g*) Digs left spade pit.

(*h*) When directed by chief of section, assisted by other cannoneers, seats spades in pits.

(*i*) Replaces left trail handspike in bracket on left trail.

(*j*) Assists ammunition corporal in the preparation of ammunition.

(*k*) Takes his post.

(11) *Ammunition corporal.* Supervises the care of ammunition and its preparation for firing (TM 9–1900).

c. If PREPARE FOR ACTION has not been ordered before the piece is established in the firing position, the command is habitually given by the chief of section as soon as the piece has been uncoupled. If this is not desired, the caution DO NOT PREPARE FOR ACTION must be given.

15. POSTS OF CANNONEERS. a. Piece coupled. See paragraph 8 and figure 7.

b. Piece uncoupled, not prepared for action. See paragraph 8 and figure 8.

c. Pieces prepared for action. See paragraph 8 and figure 9.

16. TO PREPARE TO TRAVEL (MARCH ORDER). a. General. The howitzer being uncoupled and prepared for action, to resume the order for marching, the command is: MARCH ORDER.

b. Duties of individuals prior to coupling. (1) *Chief of section.* (*a*) Supervises the work of the cannoneers.

(*b*) Verifies that the piece is not loaded.

(*c*) Verifies that traveling lock has been properly secured.

(*d*) Commands TRAILS UP when trail spades are to be removed.

(*e*) Commands CLOSE when trails are ready to be closed and guides float on left trail into brackets on right trail.

(*f*) Inspects the matériel and, when the operations have been completed, reports to the executive, "Sir, No. (so-and-so) in order," or reports any defects which the section cannot remedy promptly.

(2) *Gunner.* (*a*) Puts piece in center of traverse and elevates to approximately 400 mils.

(*b*) Sets all scales of the panoramic telescope at zero, and covers telescope mount bubbles.

(*c*) Removes telescope from mount and replaces it in its case.

(*d*) Raises top portion of left shield.

(*e*) Traverses, elevates, and depresses the piece to enable No. 5 to engage and lock the traveling lock.

(*f*) Assisted by Nos. 1 and 2, replaces rear section of howitzer cover.

(*g*) Takes his post.

(3) *No. 1.* (*a*) Inspects the chamber to see that the bore is clear and closes breech.

(*b*) After No. 5 has seated the firing mechanism, locks the percussion hammer locking pin with hammer raised.

(*c*) Replaces lanyard, vent-cleaning bit, primer-seat cleaning reamer, spare firing mechanism, waste, and oiler in section chest.

(*d*) Assists Nos. 3, 5, and 7 to raise and close right trail.

(*e*) Assisted by No. 3, replaces right trail spade in traveling position.

(*f*) Procures rear section of howitzer cover and assists gunner and No. 2 to replace it.

(*g*) Takes his post.

(4) *No. 2.* (*a*) Removes loading-rammer head, replaces it in section chest, and replaces rammer staff sections on left trail.

(*b*) Secures sledge, removes spade keys, and replaces spade keys and sledge on left trail.

(*c*) Replaces swabbing sponge in section chest and removes the pail from its position inside left trail.

(*d*) Assists Nos. 4, 6, and 8 to raise and close left trail.

(*e*) Assisted by No. 4, replaces left trail spade in traveling position.

(*f*) Assists gunner and No. 1 to replace rear section of howitzer cover.

(*g*) Takes his post.

(5) *No. 3.* (*a*) Replaces fuze setter and fuze wrench in section chest.

(*b*) Assists ammunition corporal in the preparation of ammunition for loading in prime mover.

(*c*) Assists Nos. 1, 5, and 7 to raise and close right trail.

(*d*) Assists No. 1 to replace right trail spade in traveling position.

(*e*) Takes his post.

(6) *No. 4.* (*a*) Receives jack handle from No. 5.

(*b*) Working with No. 5, returns firing jack to traveling position (Carriage M1A1).

    1. Assists No. 5 to raise piece from firing-jack key, lower piece onto wheels, and remove firing-jack float from firing-jack rack plunger.

    2. Hands jack handle to No. 5.

    3. Raises rack plunger and replaces firing-jack housing bottom cover.

    4. Raises rack plunger to traveling position and replaces firing-jack travel locking lever.

(*c*) Working with No. 5, returns firing jack to traveling position (Carriage M1A2).

1. Disengages firing-jack locking plunger from locked position.

2. Assists No. 5 to lower piece onto wheels, remove firing-jack float, raise plunger to traveling position, and secure firing-jack in traveling position.

3. Hands jack handle to No. 5.

(*d*) Replaces jack float in brackets on left trail.

(*e*) Assists Nos. 2, 6, and 8 to raise and close left trail.

(*f*) Assists No. 2 to replace left trail spade in traveling position.

(*g*) Assists Nos. 5 and 6 to replace front section of howitzer cover.

(*h*) Takes his post.

(7) *No. 5.* (*a*) After breech has been closed by No. 1, seats firing mechanism, and places spare firing mechanism on right trail.

(*b*) Prepares primers for traveling.

(*c*) Removes jack handles from right shield and hands one to No. 4.

(*d*) Working with No. 4, returns firing jack to traveling position (Carriage M1A1).

1. Assisted by No. 4, raises piece from firing-jack key.

2. Removes firing-jack key and replaces it in its carrying bracket.

3. Assisted by No. 4, lowers piece onto wheels and removes firing-jack float.

4. Raises traveling lock and connects it to the cylinder yoke, directing the gunner in manipulating the elevating and traversing handwheels.

(*e*) Working with No. 4, returns firing jack to traveling position (Carriage M1A2).

1. Manipulates ratchet so No. 4 can disengage firing-jack locking plunger from locked position.

2. Assisted by No. 4, lowers piece onto wheels, removes firing-jack float, raises plunger to traveling position, and secures jack in traveling position.

3. Replaces firing-jack locking pin.

(*f*) Replaces jack handles on right shield.

(*g*) Raises traveling lock to the traveling position, directing the gunner to traverse, elevate, or depress the piece as required to insert the lock pin, and inserts the lock pin.

(*h*) Assists Nos. 1, 3, and 7 to raise and close the right trail.

(*i*) Receives right trail handspike from No. 7, replaces it in traveling position, and latches upper strap.

(*j*) Assists Nos. 4 and 6 to replace front section of howitzer cover.

(*k*) Takes his post.

(8) *No. 6.* (*a*) Assists ammunition corporal to prepare ammunition for loading in prime mover.

(*b*) Procures and replaces aiming posts in covers. Replaces them in traveling position on right trail.

(*c*) Disassembles the rammer staff and replaces it in traveling position. Places bore brush in section chest.

(*d*) Assists Nos. 2, 4, and 8 to raise and close left trail.

(*e*) Receives left trail handspike from No. 8, replaces it in traveling position, and latches trail handspike support spring lever.

(*f*) Replaces muzzle cover on howitzer.

(*g*) Procures and, assisted by Nos. 4 and 5, replaces front section of howitzer cover.

(*h*) Takes his post.

(9) *No. 7.* (*a*) Assists ammunition corporal to prepare ammunition for loading in prime mover.

(*b*) Assisted by Nos. 1, 3, and 5, raises and closes right trail.

(*c*) Passes trail handspike to No. 5.

(*d*) Locks trail lock.

(*e*) Takes his post.

(10) *No. 8.* (*a*) Assists ammunition corporal to prepare ammunition for loading in prime mover.

(*b*) Assisted by Nos. 2, 4, and 6 raises and closes left trail.

(*c*) Passes trail handspike to No. 6.

(*d*) Replaces trail-lock pin.

(*e*) Takes his post.

(11) *Ammunition corporal.* (*a*) Supervises preparation of ammunition for return to the prime mover. Verifies that all time fuzes are set at safe (S), selective point detonating fuzes at superquick (SQ), and all booster cotter pins and safety pull wires are replaced. Determines that proper number and type increments are placed with the proper base section, before charges are returned to their containers.

(*b*) Takes his post.

c. At the completion of the above duties, the section proceeds to couple as outlined in paragraph 17. If this is not desired, the caution DO NOT COUPLE must be given.

d. To resume fire in another position. (1) If firing is to be resumed shortly in another position to which the piece must be towed by its prime mover, the command MARCH ORDER is not given. In this case, at the command for coupling, only such of the operations incident to march order are performed as are necessary for the movement of the piece and the care and security of the equipment.

(2) If the command MARCH ORDER is given while the piece is coupled, the operations pertaining to march order are completed.

17. TO COUPLE. a. The pieces being in position and in march order the command is: COUPLE. The prime movers, under the command of the first sergeant, approach from the right (left) flank. As each prime mover approaches its piece, it turns to the left (right) and halts in prolongation of the trails of the piece.

b. The gunner and all cannoneers, assisted by the ammunition corporal, under the direction of the chief of section load the unexpended ammunition, tools, and accessories. Nos. 2 to 8 inclusive hasten to the trail handles, even-numbered cannoneers on the left, odd-numbered cannoneers on the right. Gunner verifies that the drain cock on the air tank is closed. Gunner and No. 1 release left- and right-hand brakes respectively. Nos. 2 to 8 inclusive raise the trails. Gunner and No. 1 assist by placing their weight on the muzzle. The prime mover, upon signal from the chief of section, is maneuvered backward until the lunette is over the pintle. The lunette is then lowered onto the pintle. No. 7 latches and secures the pintle. Nos. 5 and 6 uncouple the air hoses from the dummy couplings on the trails of the piece and pass the hoses to Nos. 8 and 7 respectively. Nos. 7 and 8 couple the service and emergency air brake hoses, respectively, and open the service and emergency cut-out cocks on the prime mover.

CHAPTER 5

DUTIES IN FIRING

Section I. INDIRECT LAYING

18. GENERAL. In general, the duties of individuals in the howitzer sections are as follows:

a. The chief of section supervises and commands his howitzer section and is responsible that all duties of the section are performed properly, all commands executed, and all safety precautions observed.

b. The gunner sets the announced deflection and elevation, centers the cross-level bubble, lays, and refers the piece.

c. No. 1 opens and closes the breech and fires the piece.

d. No. 2 rams the projectile, assisted by No. 5.

e. No. 3 fuzes the projectile and sets the fuze.

f. No. 4 assists No. 7 to carry the projectile to the piece and places powder charge in chamber.

g. No. 5 assists No. 2 to ram the projectile; primes the piece.

h. No. 6 prepares the powder charge.

i. No. 7 carries the projectile to the piece, assisted by No. 4.

j. No. 8 inspects and cleans the projectile and assists in preparing the powder charge.

k. The ammunition corporal is responsible that ammunition is properly stored, handled, and prepared for firing.

19. CHIEF OF SECTION. a. Enumeration of duties.
(1) Assisted by the gunner, to lay for elevation when the gunner's quadrant is used.

(2) To measure the elevation.

(3) To measure the site to the mask.

(4) To indicate to the gunner the aiming point or the referring point.

(5) To follow fire commands.

(6) To indicate when the piece is ready to fire.

(7) To give the command to fire.

(8) To report errors and other unusual incidents of fire to the executive.

(9) To conduct prearranged fires.

(10) To record basic data.

(11) To observe and check frequently the functioning of the matériel.

(12) To assign duties when firing with reduced personnel.

(13) To verify the adjustment of the sighting and laying equipment.

b. Detailed description of duties. (1) *To lay for elevation when the gunner's quadrant is used.* (*a*) The command QUADRANT (SO MUCH) indicates that the gunner's quadrant is to be used.

(*b*) To set an elevation of the gunner's quadrant, for example, 361.8 mils, the chief of section sets the upper edge of the index plate opposite the 360 mark of the graduated arc on the quadrant frame and turns the micrometer on the index arm to a reading of 1.8. Care must be taken to face the same side of the quadrant in setting both the index plate and the micrometer knob.

(*c*) The announced elevation having been set on the gunner's quadrant, the piece loaded, the breechblock closed, and the cross-level bubble centered, the chief of section places the quadrant on the quadrant seat of the bracket on the telescope mount with the words "line of fire" at the bottom and the arrow pointing toward the muzzle. The chief of section must be sure to use the arrow which appears on the same side of the quadrant as the scale which he used in making the setting. He stands squarely opposite the side of the quadrant and holds it firmly on the quadrant seat, parallel to the axis of the bore. *It is important that he take the same position and hold the quadrant in the same manner for each subsequent setting, so that in each case he will view the quadrant bubble from the same angle.* If it is impracticable to use the quadrant seat of the bracket on the telescope mount, the quadrant seat on the tube will be used.

(*d*) The chief of section then causes the gunner to elevate or depress the piece until the bubble is centered, being careful that the last motion is in the direction of increasing elevation. The chief of section warns the gunner when the bubble is approaching the center, in order that the final centering may be performed accurately.

(2) *To measure the elevation.* (*a*) At the command MEASURE THE ELEVATION, the piece having been laid, the chief of section causes the gunner first to center the cross-level bubble, then to center the longitudinal-level bubble with the elevating knob. The chief of section then reads the elevation set on the elevation scale and micrometer and announces the elevation thus set; for example, "Elevation No. (so-and-so), (so much)."

(*b*) If use of the elevation scale on the telescope mount is impracticable, the chief of section may use the gunner's quadrant in measuring the elevation.

(3) *To measure the site to the mask.* (*a*) The command is: MEASURE THE SITE TO THE MASK. The chief of section, sighting along the lowest element of the bore, causes the gunner to operate the elevating handwheel until the line of sight just clears the crest at its highest point in the probable field of fire. The gunner having first centered the cross-level bubble, then centers the longitudinal-level bubble by turning the elevating knob. The chief of section reads the elevation set on the elevation scale and micrometer and reports to the executive, "Angle of site No. (so-and-so), (so much)."

(*b*) If use of the elevation scale on the telescope mount is impracticable, the chief of section may use the gunner's quadrant in measuring the site to the mask.

(*c*) When the executive announces the minimum elevation and charge, the chief of section records it in a notebook and has the gunner chalk it on a convenient place on the carriage or on the section data board.

(4) *To indicate to the gunner the aiming point or the referring point.* Whenever an aiming point or referring point has been designated by the executive, the chief of section will make sure that he has properly identified the point in question. He will then indicate it to the gunner. If there is any possibility of misunderstanding, the chief of section will turn the telescope until the horizontal and vertical hairs are on the point designated.

(5) *To follow fire commands.* The chief of section will follow the fire commands mentally. He will not repeat them, but will be prepared to give any element of the last commands to the gunner or any cannoneer who has failed to hear it.

(6) *To indicate when the piece is ready to fire.* When the executive can see arm signals or the chief of section, the chief of section will extend his right arm vertically upward as a signal that the piece is ready to fire. He gives the signal as soon as the gunner calls "Ready." When arm signals cannot be seen, the chief of section reports orally to the executive, "No. (so-and-so) ready."

(7) *To give the command to fire.* When No. 1 can see arm signals made by the chief of section, the chief of section will give the command to fire by dropping his right arm sharply to his side. When his arm signals cannot be seen, he commands orally NO. (SO-AND-SO) FIRE. The chief of section will not give the signal or command to fire until all the cannoneers are in their proper places. He will require the cannoneers to stand clear of the piece until it is firmly seated.

(8) *To report errors and other unusual incidents of fire to the executive.* If for any reason the piece cannot be fired, the chief of section will report promptly that fact to the executive, and the reason therefor; for example, "No. (so-and-so) out, misfire." Whenever it is discovered that the piece has been fired with an error in laying, the chief of section will report that fact at once; for example, "No. (so-and-so) fired 40 mils right." Whenever the gunner reports that the aiming posts are out of alinement with the sight, the chief of section will report that fact and request instructions. Likewise, he promptly reports other unusual incidents that affect the service of the piece. See paragraph 44.

(9) *To conduct prearranged fires.* Whenever the execution of prearranged fires is ordered, the chief

of section will conduct the fire of his section in strict conformity with the prescribed data.

(10) *To record basic data.* The chief of section will record data of a semipermanent nature in a notebook. These include such data as minimum elevation; base deflection, including aiming points used; prearranged fires, when prepared schedules are not furnished; safety limits in elevation and deflection; number of rounds fired, with the date and hour; and calibration corrections when appropriate.

(11) *To observe and verify the functioning of matériel.* The chief of section closely observes the functioning of all parts of the matériel during firing. Before the piece is fired, he verifies that the recoil mechanism contains the proper amount of oil; thereafter he carefully observes the functioning of the recoil system. He promptly reports to the executive any evidence of malfunctioning.

(12) *To assign duties when firing with reduced personnel.* Whenever the personnel of the section serving the piece is temporarily reduced in numbers below that indicated in this manual, the chief of section will make such redistribution of duties as will best facilitate the service of the piece.

(13) *To verify adjustment of sighting and laying equipment.* See appendix II and TM 9–331.

20. GUNNER. a. Enumeration of duties. (1) To set or change the deflection.

(2) To apply deflection difference.

(3) To set the elevation.

(4) To center the cross-level bubble.

(5) To lay for direction.

(6) To lay for elevation.

(7) To call "Ready."

(8) To refer the piece.

(9) To record base deflection.

(10) To measure a deflection.

b. Detailed description of duties. (1) *To set or change the deflection.* (*a*) *To set deflection.* At the command, for example, DEFLECTION 483, the gunner first sets the azimuth micrometer index (movable) to its zero position and sets the azimuth micrometer at zero. Using his left hand, he pushes the throw-out lever, and with his right hand turns the rotating head until the numeral 4 on the azimuth scale appears opposite the azimuth scale index. He then grasps the azimuth worm knob with his right thumb and forefinger and turns the knob clockwise until the numeral 83 on the azimuth micrometer appears opposite the azimuth micrometer index (movable). The line of sight now makes a horizontal angle of 483 mils with the 0–3200 line of the panoramic telescope. The gunner now turns the azimuth micrometer index (movable) opposite an even ten graduation on the micrometer in preparation for setting off the next shift. This last movement does not change the setting of the azimuth scales.

(*b*) *To change deflection.* The gunner should be trained to always grasp the azimuth worm knob with his right thumb and forefinger. He also should be taught that moving his thumb upward (clockwise) will cause the deflection to increase, and the tube must be traversed to the left to bring the line of sight back on the aiming point or aiming posts. Similarly, he should be taught that moving his thumb downward (counterclockwise) causes the deflection to decrease and results in a right shift in the tube when the weapon is relaid. The deflection having been set at 483 mils, if a subsequent shift of RIGHT 55 is commanded, the gunner moves his right thumb downward (counterclockwise) on the azimuth worm knob until the deflection is decreased 55 mils. If the gunner had set the azimuth micrometer index (movable) opposite 80 on the micrometer before the deflection change was given, the numeral 25 would appear opposite the azimuth micrometer index (movable) after the change. However, the true reading on the azimuth scale is obtained by turning the movable index back to its zero position and noting the reading on the scales, in this case (483 mils − 55 mils) 428 mils. The azimuth micrometer index (movable) permits the gunner to start from an even ten graduation each time a shift is given. The gunner, having set off RIGHT 55, would move the azimuth micrometer index (movable) opposite 20 or 30 on the micrometer in preparation for the next shift. Should the command be LEFT (SO MUCH), the gunner changes the setting by moving his thumb upward (clockwise) on the azimuth worm knob, thus increasing the deflection.

(2) *To apply deflection difference.* (*a*) The command is: ON NO. (SO-AND-SO) OPEN (CLOSE) (SO MUCH). The gunner of the piece indicated in the command does not change the deflection set on his telescope. Each of the other gunners changes his deflection setting by the number of mils specified in the command if his piece is next in line to the piece indicated; by twice this number of mils if his piece is second in line from the piece indicated; by three times this number of mils if his piece is third in line from the piece indicated; and so forth.

(*b*) If the command is, for example, ON NO. 2 OPEN 5, the gunner on No. 2 makes no change;

the gunner on No. 1 turns the azimuth worm knob by moving his right thumb downward (counterclockwise) to decrease his setting 5 mils; the gunner on No. 3 moves his thumb upward (clockwise) to increase his setting 5 mils; and the gunner on No. 4 moves his thumb upward until the deflection is increased 10 mils (twice the number of mils specified in the command).

(*c*) Should the command be, for example, ON NO. 3 CLOSE 10, the gunner on No. 3 does not change his setting; the gunner on No. 1 moves his right thumb upward until his deflection setting has been increased 20 mils; the gunner on No. 2 moves his thumb upward until 10 mils have been added to the deflection; and the gunner on No. 4 moves his thumb downward until the setting has been decreased 10 mils.

(*d*) When a deflection change and a deflection difference are announced at the same time (for example, RIGHT 30, ON NO. 1 CLOSE 5), both of which affect the gunner's piece, he should first set off the deflection change and then apply the deflection difference.

(*e*) In the methods described above, it is implied that the gunner resets the azimuth micrometer index (movable) opposite an even ten graduation each time the azimuth worm knob has been turned. This facilitates setting off the tens and units on the azimuth micrometer scales. The gunner, before turning the azimuth worm knob, should verify that the movable index coincides exactly with the even ten graduation he has chosen.

(3) *To set elevation.* The gunner is first taught to read elevations set and then to set announced

elevations. The telescope mount is provided with an elevation scale graduated in hundreds of mils from zero to 1,100 mils and a micrometer graduated in mils from zero to 100. The scale is read opposite the elevation scale index and the micrometer opposite the micrometer index. To set an announced elevation, the gunner grasps the elevation knob and rotates it in the appropriate direction until the announced elevation is read on the elevation and micrometer scales opposite their appropriate indexes. The last motion should always be in the direction of increasing elevation.

(4) *To center cross-level bubble.* The gunner operates the cross-leveling knob until the bubble is centered.

(5) *To lay for direction.* (*a*) The deflection having been set, the deflection difference applied (if applicable), and the cross-level bubble centered, the gunner brings the vertical hair of the panoramic telescope on the aiming point by traversing the piece. He verifies the centering of the longitudinal-level and cross-level bubbles and re-lays if necessary.

(*b*) *Procedure to insure accuracy.* To take up lost motion, the final movement of the traversing handwheel should be such as to cause the vertical hair of the telescope to approach the aiming point or target from the left. The gunner should habitually lay with the vertical hair of the telescope on exactly the same portion of the aiming point and insure that the cross-level bubble is centered for each round.

(6) *To lay for elevation.* (*a*) *With the elevation quadrant on the telescope mount.* The elevation having been set, the gunner turns the elevating handwheel until the longitudinal-level bubble is centered.

(*b*) *With the gunner's quadrant.* See paragraph 19*b*(1).

(*c*) *Procedure to insure accuracy.* In order to insure accuracy in laying for elevation, the last movement of the elevating handwheel must be in the direction of increasing elevation. When firing at elevations in excess of 250 mils, after each round the gunner will depress the piece to approximately 250 mils to facilitate loading.

(7) *To call "Ready."* The piece having been laid for direction and elevation and No. 1 having called "Set," the gunner verifies the laying, moves his head clear of the telescope, and calls "Ready" to indicate that the piece is ready to be fired.

(8) *To refer the piece.* The piece having been laid for direction, to refer the piece, the command is: AIMING POINT (SO-AND-SO), REFER. Without disturbing the laying of the piece, the gunner brings the vertical hair of the telescope on the new aiming point (referring point). He then reads and announces the deflection thus set. The piece may be laid subsequently using this referring point as an aiming point. This method affords a convenient means of laying when the initial aiming point is either not convenient or not permanent. The piece is normally referred to the aiming posts, but should also be referred to one or more distant aiming points for accuracy and permanence. When so ordered, the chief of section records the deflection and description of each referring point in his notebook. The gunner records the deflection and aiming point in current use on a convenient part of the piece.

(9) *To record base deflection.* At the command RECORD BASE DEFLECTION, the gunner re-

cords the deflection set on his telescope upon some convenient part of the shield or upon a data board.

(10) *To measure a deflection.* The command is: AIMING POINT (SO-AND-SO), MEASURE THE DEFLECTION. The piece having been established in direction, the gunner turns the telescope until the vertical hair is on the aiming point. He then reads and announces the deflection.

21. No. 1.   a. Enumeration of duties.   (1) To open and close the breech.

(2) To operate the percussion hammer locking-pin knob so as to lock and unlock the percussion hammer.

(3) To attach and detach the lanyard.

(4) To clean and oil the breechblock and the breech recess.

(5) To clean the primer vent and the primer seat.

(6) To call "Set."

(7) To fire the piece.

b. Detailed description of certain duties.   (1) *To open the breech.* After the firing mechanism has been removed by No. 5, No. 1 slides the breech operating lever closing latch to the left, pulls the operating lever downward to the horizontal position, and swings the lever to the right until the breech is fully open. No. 1 then locks the percussion hammer by means of the locking-pin knob and detaches the lanyard.

(2) *To close the breech.* No. 1 grasps the breech operating lever, swings it backward and to the left until the breech is fully closed and then upward until it is engaged by the breech operating lever closing latch. After No. 5 has seated the firing mechanism and the gunner has called "Ready," No.

1 attaches the lanyard and unlocks the percussion hammer. *No. 1 will not attach the lanyard until the gunner has called "Ready."* The chief of section may caution, "With the long lanyard," in which case No. 1 attaches the long lanyard.

(3) *To clean and oil the breechblock and the breech recess.* Immediately upon opening the breech after a round has been fired, No. 1 will wash the powder residue from the mushroom head, the gas check seat, and the threaded sectors of the breech recess and breechblock with a water-saturated cloth, then oil the parts with a cloth *slightly dampened* with oil, lubricating, preservative, medium (for temperatures of 0° F. and above) or oil lubricating, preservative, special (below 0° F.). When necessary, he will oil the operating parts of the breech mechanism with the same oil as specified for the breech recess and breech-block.

(4) *To clean the primer vent and primer seat.* Upon completion of the duty prescribed in (3) above, No. 1 cleans the primer vent with the vent cleaning bit and removes residue from the primer seat with the primer-seat cleaning reamer.

(5) *To call "Set."* The breech having been closed and the firing mechanism fully seated, No. 1 calls, "Set."

(6) *To fire the piece.* At the chief of section's signal or command NO. (SO-AND-SO) FIRE (par. 19**b**(7)), No. 1 grasps the handle of the lanyard with his right hand and pulls strongly with a quick movement to the right rear, prolonged sufficiently to insure that the hammer hits the firing pin. In case of a misfire, the instructions contained in paragraph 49 will be followed.

22. NO. 2.   a. Enumeration of duties.   (1) To swab and inspect the powder chamber after each round.

(2) To ram the projectile, assisted by No. 5.

(3) To call out the number of the round and the announced elevation in volley fire.

(4) To swab out the bore, assisted by No. 4.

b. Detailed description of duties.   (1) *To swab and inspect the powder chamber after each round.* After each round, No. 2 sponges out the powder chamber immediately after No. 1 opens the breech. The rear of the bore, including the forcing cone, is swabbed with a sponge dipped in water. Before the piece is loaded, and after swabbing between rounds, No. 2 inspects the bore for injuries to the tube, for burning fragments of powder bags or other objects. Any burning fragments and other objects in the bore must be removed before reloading. Any injury to the howitzer will be reported to the chief of section.

(2) *To ram the projectile.*   As soon as the lip of the loading tray is placed in the breech recess, No. 2 procures the rammer staff and, working on the left, places the rammer head against the base of the projectile. Assisted by No. 5 on the right, No. 2 pushes the projectile into the breech recess until the base of the projectile clears the rear of the powder chamber (fig. 16). No. 2 then commands READY, RAM. At the second command, Nos. 2 and 5 drive the projectile forward into the forcing cone using their arms and weight to add power to the stroke (fig. 17). *Uniformity of ramming is absolutely essential to accuracy of fire.* Firm seating of the projectile is necessary to prevent it from slipping back into the powder chamber and resting on the charge,

Figure 16. Pushing projectile from tray into breech recess.

53

Figure 17. Ramming the projectile.

especially at high elevations. After the projectile is rammed, No. 2 returns the rammer staff to its position near the left trail.

(3) *To call out the number of the round and the announced elevation in volley fire.* To insure that the correct number of rounds is fired in volley fire, No. 2 calls out the number of the round and the elevation as he finishes ramming the projectile. As he finishes ramming the last round, he adds, "Last round." For example, when two rounds are to be fired at 480, he calls out, "Second and last round, 480." He should not speak louder than is necessary to insure his being heard by the members of his own section.

(4) *To swab out the bore.* The chief of section directs the swabbing out of the bore during intervals of the firing or at other times. The howitzer is brought to the horizontal by the gunner. No. 2, assisted by No. 4, swabs out the bore as prescribed in TM 9–331.

23. NO. 3.   a. Enumeration of duties. (1) To fuze projectiles.

(2) To set the fuze setter.

(3) To set fuzes.

(4) To remove fuzes from projectiles.

b. Detailed description of duties. (1) *To fuze projectiles.* No. 3 unscrews the eyebolt lifting plug from the fuze socket of the projectile and inspects the socket for rust and dirt. If the VT fuze is designated he removes the supplemental charge, if present, and screws the fuze home by hand. If selective superquick and delay, combination time and superquick, concrete piercing or mechanical time fuze is designated, he insures that the supplemental charge is present, provided this type of projectile is used; removes the booster cotter pin from the designated fuze, and screws the fuze home by hand. In the case of the concrete piercing fuze with separate booster, the booster cotter pin is removed and the booster is screwed into the projectile by means of the wrench supplied for that purpose; the fuze is then screwed into the projectile. The fuze is given its final seating by the use of the fuze wrench (fig. 20). No great force should be used. If there is any difficulty in screwing the fuze home, and the threads are clean, the fuze should be removed and another inserted. If the same trouble is encountered with the second fuze, the projectile should be rejected. If a time fuze is to be used the safety pull wire or cotter pin is then removed.

(2) *To set the fuze setter.* (a) *Using fuze setter M22 or M23.* No. 3 releases the corrector scale clamping screw marked "C," and grasping the handle, turns the body and time scale until the index on the time scale is opposite the announced corrector setting on the corrector scale. He clamps the corrector scale clamping screw, being careful not to disturb the corrector setting. He then releases the time scale clamping screw marked "T," and grasping the handle, turns the body until the index on the body is opposite the announced time on the time scale. He then locks the time scale clamping screw, being careful not to disturb the setting. For accuracy, No. 3 locks squarely at the scales and indexes in the same manner each time.

(b) *Using fuze setter M14.* This is a wrench type fuze setter which does not employ time or corrector scales.

(3) *To set fuzes.* (a) *Selective superquick and*

delay fuzes. When **FUZE QUICK** is announced, No. 3 will verify the superquick setting. (The slot on the setting sleeve should be alined with the letters **SQ**.) When **FUZE DELAY** is announced, he will turn the setting sleeve until the slot is alined with the word **DELAY**.

(*b*) *Combination time and superquick fuzes.* These fuzes may be set for time action. However, the percussion element will detonate the round upon impact if the time element fails. After fuzing the projectile No. 3 removes the safety pull wire from the fuze. The method of setting for time action is given in (*e*) below. For percussion action the command is: **FUZE QUICK M (SO-AND-SO)**. No. 3 verifies that the "S" on the setting ring is alined with the index on the fixed ring. If not, he so sets it.

(*c*) *Mechanical time fuzes.* These fuzes contain no percussion element; therefore, they must be set to function at some time less than the time of flight of the projectile if a high order burst is to be expected. For the method of setting the desired time see (*e*) below.

(*d*) *VT fuzes.* These fuzes operate and function in such a manner as to require no setting on the part of any personnel.

(*e*) *Setting time fuzes.* See (*b*) and (*c*) above.

　　1. *Using fuze setter M22 or M23.* After making the announced settings on the fuze setter, No. 3 carefully places it over the fuze and turns the setter clockwise until the notch on the time ring of the fuze engages the stop on the setting ring of the fuze setter. He then turns the handle to the horizontal position, pushes down on the fuze setter until the notch

fully engages the stop, and continues to turn it clockwise 'until the pawl in the adjusting ring assembly drops into the notch of the fixed fuze ring. This prevents further turning and indicates that the fuze is set. He then lifts the fuze setter from the fuze.

2. Using fuze setter M14. No. 3 engages the key on this wrench type fuze setter in the notch on the setting ring and rotates the setting ring until the announced time setting is opposite the index on the fixed ring. When using the fuze setter M14, the commands from the battery executive do not include corrector as he determines the required time setting from the data received from the officer conducting fire.

(4) *To remove fuzes from projectiles.* If for any reason a projectile which has been fuzed is not to be fired, the fuze will be removed. The operation of inserting a fuze is reversed. Supplemental charges will be replaced, provided the projectile was issued with the charge. The booster cotter pin of the fuze is replaced, if provided with the fuze. Combination superquick and delay fuzes are reset to superquick. Time fuzes are reset to S (safe) and the safety pull wire replaced. The eyebolt lifting plugs are replaced in the fuze sockets of the projectiles and all fuzes returned to their containers.

24. NO. 4. a. Enumeration of duties. (1) To assist No. 7 to carry projectile to the piece.

(2) To place powder charge in chamber.

(3) To assist No. 2 to swab out the bore.

b. Detailed description of duties. (1) *To assist*

Figure 18. Inserting powder charge in chamber.

Note.—Loading rammer head is being placed on ammunition or fuze box cover to protect it from the dirt.

No. 7 to carry projectile to the piece. No. 4 places the fuzed projectile on the loading tray and grasps the handles on the left side of the tray. No. 7 grasps the handles on the right side, and together they carry the tray to the piece. They place the lip of the tray in the breech recess against the rear face of the tube and hold it there until Nos. 2 and 5 push the projectile into the chamber with the rammer. No. 4 then releases the handles and the tray is then withdrawn by No. 7.

(2) *To place powder charge in chamber.* After releasing the tray, No. 4 turns and receives the powder charge from No. 6. He places the powder charge in the chamber, lashed end to the front, and pushes it in until the base of the charge is flush with the rear of the chamber (fig. 18). To insure transmission of the flash from the primer to the charge when the breech is closed, the mushroom-head (obturator) should come in contact with the base of the charge, push it forward to its final position, and remain in contact with it, with the igniter pad directly in front of the vent. No. 4 must be sure that the red igniter pad is to the rear and that the igniter protector cap has been removed.

(3) *To assist No. 2 to swab out the bore.* See paragraph 22**b**(4).

25. NO. 5. a. Enumeration of duties. (1) To prime the piece.

(2) To assist No. 2 to ram the projectile.

(3) To clean and oil the firing mechanism M1, when used.

b. Detailed description of duties. (1) *To prime the piece.* Immediately after the piece returns to battery after a round is fired, No. 5 disengages the

Figure 19. Priming the piece.

firing mechanism safety latch and unscrews the firing mechanism by turning it counterclockwise. After No. 1 has fully closed the breech No. 5 inserts the alternate primed firing mechanism in the housing, taking care that the front end of the primer has entered the obturator spindle plug. He then seats the firing mechanism by turning the handle clockwise until it contacts the firing mechanism block handle arm stop and is latched (fig. 19). It is important to make sure that the firing mechanism is screwed home and latched in position, as it is possible to fire the piece even though the firing mechanism is not completely in its proper firing position. If the piece is fired without this having been done, damage to the breech-block and injury to personnel may result. Should a primer be slightly oversize, or the primer seat dirty, the mechanism will stick before it has fully seated. Force should not be exerted to seat the firing mechanism. It should be removed, the primer seat cleaned, or another primer inserted. Unfired primers to be discarded are turned over to the ammunition corporal for disposal.

(2) *To assist No. 2 to ram the projectile.* See paragraph 22**b**(2).

(3) *To clean and oil the firing mechanism.* Two firing mechanisms are provided for each piece, and they should be used alternately. As soon as one firing mechanism is inserted in the breechblock, No. 5 will remove the fired primer from the other firing mechanism, wipe it, and insert a new primer for use with the next round. He holds the firing mechanism in his left hand with the notch in the primer holder facing up while he slides the base of the expended primer out from the holder. In seating a new primer in the primer holder he holds the firing mechanism in

Figure 20. Preparing ammunition.

63

the same manner, taking the precaution to keep his right hand clear of the front end of the primer. During lulls in the firing, both firing mechanisms will be cleaned and oiled thoroughly.

26. NO. 6.   a. Enumeration of duties.   (1) To prepare powder charges.

(2) To pass powder charge to No. 4.

(3) To call out the number of the charge.

b. Detailed description of duties.   (1) *To prepare powder charges (fig. 20)*.   (*a*) *Type M3 (green bag)*. This propelling charge is a multiple-section charge of smokeless powder with the necessary black-powder igniter, and consists of a base charge and four smaller increments corresponding to five zones of fire. The base charge is marked charge 1 and the other bags are numbered from 2 to 5, inclusive. When the command designating the charge to be used is given, for example, CHARGE 3, No. 6 takes a complete charge from one of the containers, places the complete charge in front of him, base charge on the bottom, and unties the lashings which hold the bags together. Without disturbing the order in which they are arranged, he checks the increments, removes the bags marked 4 and 5 from the top, leaving the bag marked 3 at the top of the pile. He then ties the remaining bags together, removes the igniter protector cap on the base of the charge, and passes the charge thus prepared to No. 4. The discarded bags are placed in the containers provided for that purpose, and disposed of later as the executive may direct.

(b) *Type M4A1 (white bag)*. Type M4A1 (white bag) is composed of a base charge, with igniter, and four increments corresponding to charges 3, 4, 5, 6, and 7, respectively. The preparation of a charge with this type is similar to that described in (*a*) above, except that no charge lower than charge 3 can be prepared. *Under no circumstances will green and white bag increments be mixed in the same charge.*

(2) *To pass powder charge to No. 4.* See figure 17.

(3) *To call out the number of the charge.* After handing the charge to No. 4, No. 6 calls out the number of the charge which he has prepared to verify with his chief of section that the proper charge has been prepared.

27. NO. 7. a. Enumeration of duties. (1) Assisted by No. 4, to carry projectiles to piece.

(2) To assist No. 6 in preparing powder charges.

b. Detailed description of duties. (1) *Assisted by No. 4, to carry projectiles to the piece (figs. 16 and 20).* After No. 4 has placed the fuzed projectile on the loading tray and grasped the handles on the left side of the tray, No. 7 grasps the handles on the right side, and together they carry the tray to the piece. They place the lip of the tray in the breech recess against the rear face of the tube and hold it there until Nos. 2 and 5 push the projectile into the chamber with the rammer. No. 4 then releases his handles and No. 7 returns the tray to the place where the ammunition is being prepared.

(2) *To assist No. 6 to prepare powder charges.* No. 7 opens powder containers. He assists No. 6 to untie and make up powder charges.

28. NO. 8. a. Enumeration of duties. (1) To inspect and clean projectiles.

(2) To hold projectile upright while it is being fuzed and the fuze is being set by No. 3 (fig. 20).

b. Detailed description of duties. (1) *To inspect and clean projectiles.* No. 8 verifies type, weight, and lot numbers (when appropriate) of each projectile, removes the grommet, and examines the projectile carefully for defects. The rotating bands will be inspected with care; if any burrs are found they will be removed with a file. The projectile is then placed upright on its base and the entire surface cleaned with a piece of waste or, if necessary, with a sponge and water. Should any appreciable length of time intervene between the cleaning of the projectile and its insertion into the piece, the projectile must be reinspected before loading to see that it is free from sand and dirt. Any sand or dirt on the projectile will cause undue wear, scratches, or gouges in the bore of the weapon.

(2) *To hold projectile upright while it is being fuzed and the fuze is being set.* After the command which designates the projectile to be used is given, for example, SHELL HE, No. 8 gets the projectile and holds it upright while No. 3 fuzes it and sets the fuze. When directed by the chief of section, No. 8 will read and announce the time as set.

29. AMMUNITION CORPORAL. a. Enumeration of duties.
(1) To receive and account for ammunition for the section.

(3) To enforce proper methods of handling ammunition.

(3) To supervise the storage of ammunition.

(4) To have ammunition properly prepared for firing.

(5) To insure that the designated powder charge, projectile, and fuze are used.

b. Detailed description of duties. (1) *To receive and account for ammunition for the section.* Subject

Figure 21. Piece loaded and ready to fire.

to the orders of the executive of the chief of section, the ammunition corporal will obtain such ammunition as may be required by the section from the battery ammunition dump or the battery ammunition vehicles. He will verify the amount received and receipt for it. He will maintain a daily record of all ammunition received and fired. He will keep the chief of section informed as to the status of the ammunition supply within the section.

(2) *To enforce proper methods of handling ammunition.* The ammunition corporal will require the cannoneers to handle ammunition properly. He will prevent any of the following:

(*a*) Smoking by anyone handling, or in the vicinity of, ammunition.

(*b*) Use of any lights, other than flashlights, in the vicinity of powder charges.

(*c*) Dropping projectiles, powder containers, fuzes, and primer boxes from vehicles.

(*d*) Allowing projectiles to strike together.

(*e*) Allowing ammunition to become dirty, wet, or overheated.

(3) *To supervise the storage of ammunition.* See paragraph 50.

(4) *To have ammunition properly prepared for firing.* The ammunition corporal will carefully supervise the work of the cannoneers in preparing rounds for firing. He will see that projectiles are cleaned thoroughly and that all burrs on the rotating bands have been removed by filing. He will require all powder charges to be kept in their containers until just before loading. He will require lids of powder containers to be kept closed except when a charge is being withdrawn, and primers and fuzes to be kept in their boxes until just before using. He will insure

that powder charges are properly segregated by lot number, and that increments do not become mixed.

(5) *To insure that the designated powder charge, projectile, and fuze are used.* The ammunition corporal will follow the fire commands and will indicate, when necessary, to the cannoneers concerned the projectiles, powder charges, and fuzes to be used. For any single firing mission, he will see that the projectiles are all of one weight and that the powder charges and time fuzes are all of one lot number.

Section II. DIRECT LAYING

30. GENERAL. Delivery of fire by direct laying demands a high degree of training in its special techniques since it requires the section to operate as an independent unit. This training is based on the technique employed in the normal mission of indirect laying. Since the targets involved in direct laying are usually capable of firing on the howitzer section at point-blank range, the high standards of speed and accuracy required in indirect laying become even more important for direct laying missions.

31. PREPARATION OF A RANGE CARD. In any position, at the earliest possible time, the chief of section measures or estimates the range to critical points in likely avenues of approach for enemy vehicles and tanks, converts the range to the corresponding elevations for charge 7, and makes a range card (fig. 22) with the ranges and corresponding elevations for quick reference. If no prominent terrain features are available, stakes may be driven in the ground for range reference points. As time permits, the range card will be improved by noting ranges obtained by firing, pacing, taping, vehicle odometer reading, measurements made from a map, or survey.

32. FIELD OF FIRE. The sector of fire for the piece should, if possible, be cleared of all obstructions which will endanger battery personnel when the piece is fired.

33. CHIEF OF SECTION. a. Enumeration of duties. (1) To conduct fire of his piece.

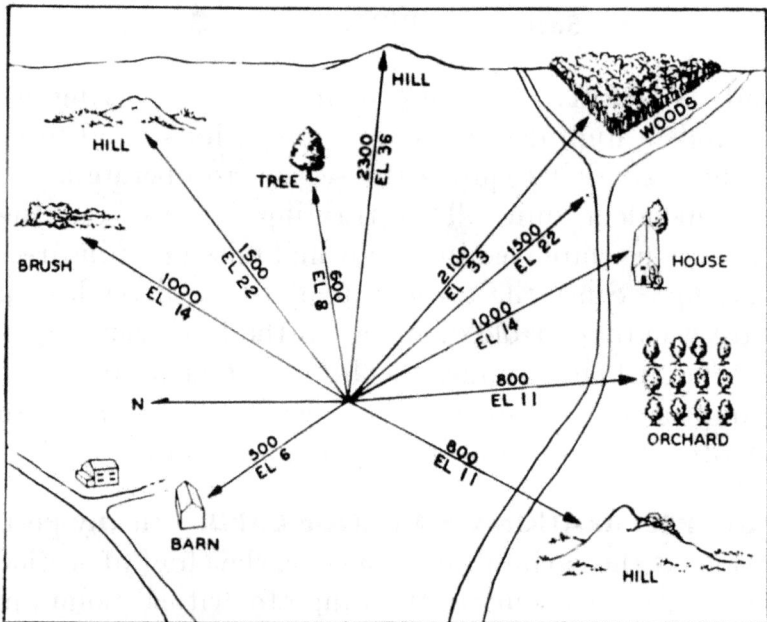

Figure 22. Range card for direct laying.

(2) To identify or select the target.
(3) To estimate the range to the target.
(4) To determine the lead in mils.
(5) To give initial commands.
(6) To give subsequent commands, based on observed effect.

b. Detailed description of certain duties. (1) *To conduct the fire of his piece.* The chief of section conducts the fire of his piece when the executive commands TARGET (IDENTIFICATION), FIRE AT WILL, or simply FIRE AT WILL.

(2) *To identify or select the target.* If the executive designates an object or one of a group of objects as the target, the chief of section must correctly identify this target. If the target is a group of tanks or other objects, the chief of section selects the target which, in his estimation, constitutes the greatest danger to his own position or the position of the supported troops.

(3) *To estimate the range to the target.* A range card with accurately measured ranges to key points provides the best means for determining the initial range. If a range card has not been prepared the range is estimated.

(4) *To determine the lead in mils.* The appropriate lead in mils for targets moving at various speeds, when HE with charge 7 is fired, is as follows:

Lateral speed	*Lead*
Under 10 mph	5 mils
Over 10 mph	10 mils

(5) *To give initial commands.* The chief of section will give fire commands containing the following elements in sequence:

(*a*) *Designation of target.* The command is: TARGET (SO-AND-SO).

(*b*) *Projectile, charge, and fuze.* Shell HE with charge 7 will be used against all types of direct laying targets. Use DELAY setting on the fuze in direct laying except against unarmored personnel. However, if less than 50 percent of the rounds fired during adjustment are ricochet bursts, the fuze setting should be changed to superquick. The M78 concrete-piercing fuze should be used against concrete pill-boxes or fortifications.

(c) *Lead.* The command is: LEAD (SO MUCH). See (4) above for method of determining lead.

(d) *Method of fire.* Fire is continuous unless otherwise commanded.

(e) *Elevation.* The command is: ELEVATION (SO MUCH). See (3) above.

Note. Commands in (b) and (d) above may be omitted when the state of training of the section or standing operating procedure permit.

(6) *To give subsequent commands, based on observed effect.* (a) *Change in lead.* During adjustment the lead is changed by the command RIGHT (LEFT) (SO MUCH).

(b) *Change in elevation.* During adjustment the elevation is increased by the command ADD (SO MUCH) and decreased by the command DROP (SO MUCH). See paragraph 39 for the method of determining changes in elevation during adjustment of fire.

34. GUNNER. a. Enumeration of duties. (1) To set the elevation indexes of the panoramic telescope at zero.

(2) To set the azimuth scale and micrometer at zero.

(3) To lay on the target with the announced lead.

(4) To track the target with the traversing handwheel.

(5) To operate the elevating handwheel to keep the horizontal line in the reticle on the target.

(6) To command FIRE.

(7) To re-lay on the target.

(8) To follow subsequent commands.

b. Detailed description of duties. (1) *To set the elevation indexes of the panoramic telescope at zero.* The gunner sets the elevation index and the elevation micrometer of the panoramic telescope at zero.

(2) *To set the azimuth scale and micrometer at zero.* The gunner sets the azimuth scale and micrometer at zero.

(3) *To lay on target with announced lead and track target with traversing and elevating handwheels.* The gunner tracks the target with the traversing and elevating handwheels, keeping the vertical hair of the panoramic telescope ahead of the target by measuring the announced lead on the reticle scale, and keeping the horizontal line of the reticle on the center of mass of the target.

(4) *To command FIRE.* After No. 1 and No. 3 call "Set," and "Ready," respectively, and when ready, the gunner commands **FIRE**.

(5) *To re-lay on target and follow subsequent commands.* The gunner immediately re-lays on the target, and if a change in lead is announced, tracks the target as before, using the new lead.

35. NO. 1. a. Enumeration of duties. See paragraph 21a.

b. Detailed description of certain duties. (1) See paragraph 21b(1) to (5), inclusive.

(2) *To fire the piece.* At the gunner's command FIRE, No. 1 fires the piece in the same manner as prescribed for indirect laying (par. 21b(6)).

36. NO. 3. a. Enumeration of duties. (1) To take position to the left of the piece (fig. 23).

Figure 23. Serving the piece in direct laying.

(2) To set announced elevation.

(3) To cross-level the sight mount.

(4) To call "Ready."

b. Detailed description of duties. (1) *To take position to the left of the piece.* No. 3 places himself to the left of the piece, outside of the trail, at a point where he can operate the elevation knob and the cross-leveling knob of the sight mount, without interfering with the gunner.

(2) *To set the announced elevation.* No. 3 grasps the elevation knob and rotates it in the appropriate direction until the announced elevation is read on the elevation and micrometer scales, opposite their appropriate indexes.

(3) *To cross-level the sight mount.* No. 3 operates the cross-leveling knob until the bubble is centered.

(4) *To call "Ready."* No. 3 calls "Ready" when he has completed the duties given in (2) and (3) above.

37. AMMUNITION CORPORAL. a Enumeration of duties.
(1) To fuze projectiles.

(2) To set fuzes.

b. Description of duties. (See paragraph 23**b**(1) and (3).

38. REMAINDER OF HOWITZER SECTION.
The remaining cannoneers perform their duties as prescribed in indirect laying.

39. CONDUCT OF FIRE. a. Trajectory characteristics.
When charge 7 is fired, the following trajectory characteristics govern the manner of conducting fire.

(1) *Ranges from 0 to 400 yards.* Within these range limits the trajectory will be too flat to permit

an 8-foot tank to pass under it. The upper limit of 400 yards is the ideal at which to open fire at an approaching tank, since fire can then be conducted without misses if deflection is correct.

(2) *Ranges from 400 to 1,500 yards.* These range limits include the zone in which the trajectory is sufficiently flat to permit direct estimation of range errors without actually bracketing the target. Assuming zero dispersion, if a hit is obtained at the bottom of an 8-foot tank at the upper (1,500-yard) limit, a 100-yard range change will result in a hit at the top of the tank. During adjustment within this zone, range changes should seldom be more than 100 yards and frequently changes of 50 yards will be sufficient. (A 2-mil and 1-mil change in elevation corresponds approximately to a 100-yard and 50-yard change in range, respectively. The upper limit is the greater range at which fire should be opened unless tactical conditions require otherwise. The second shot (or certainly the third) should be a hit.

(3) *Ranges from 1,500 to 2,500 yards.* The ranges from 1,500 to 2,500 yards represent a zone in which hits are reasonably possible. Ordinarily, bracket methods are used to obtain an adjustment in this zone. Dispersion is a considerable factor in this zone; fire should not be opened at these ranges unless surprise is of no consideration.

(4) *Ranges over 2,500 yards.* At ranges above 2,500 yards, direct laying is not advisable against moving targets. Dispersion is the controlling factor. Ranges must be known accurately or determined by bracketing. At ranges over 2,500 yards, the slope of fall of the projectile becomes so great that a hit is very difficult to obtain on a moving target.

(5) *Elevation changes.* For ranges up to 1,500 yards an elevation change of 1.5 mils changes the range approximately 100 yards. A 1-mil change in elevation raises or lowers the path of the projectile (trajectory) as follows: 1.5 feet at 500 yards range, 3.0 feet at 1,000 yards range, and 4.5 feet at 1,500 yards range.

b. Vertical displacement table. The following table shows vertical displacement for 100-yard range change, 155-mm howitzer M1, firing shell HE, charge 7.

Range (yds)	Vertical Displacement (ft)	Range (yds)	Vertical Displacement (ft)
100	0.5	1400	6.5
200	1.0	1500*	7.0
300	1.5	1600	7.5
400*	1.5	1700	9.0
500	2.0	1800	9.5
600	2.5	1900	10.5
700	3.0	2000	11.0
800	3.5	2100	11.5
900	4.5	2200	12.0
1000	5.0	2300	12.5
1100	5.5	2400	13.0
1200	6.0	2500*	15.0
1300	6.0		

*Critical direct laying ranges.

CHAPTER 6

ADDITIONAL INFORMATION ON SERVICE OF THE PIECE AND SAFETY PRECAUTIONS

---

40. ACCURACY IN LAYING. Sighting and laying instruments, fuze setters, and elevating and traversing mechanisms will be manipulated to minimize the effects of lost motion. This requires that the last motions in setting instruments and in laying be always in the directions prescribed. To insure accurate laying, the gunner and any other cannoneers who have duties in connection with laying the piece will invariably be required to verify the laying after the breech has been closed.

41. AIMING POSTS. **a.** After the piece has been initially laid for direction, if a suitable natural aiming point is not visible, the piece is referred to the aiming posts as described in paragraph 20**b**(8). Two aiming posts are used for each piece. Each post is equipped with a light for use at night. One post is set up at least 100 yards from the piece. The other post is set up at the midpoint between the first post and the piece, and is lined in by the gunner so that the vertical hair of the telescope and the two aiming posts are all in the same vertical plane. It is imperative that the far post be placed at twice the distance to the near post. Any displacement of the piece during firing can be easily detected and corrected as indicated in paragraph 42. For night use, the lights should be adjusted so that the far one will appear several feet above the near one. Two lights thus will clearly establish a vertical line on which the vertical hair of the telescope can be laid.

b. The panoramic telescope is mounted a considerable distance away from the center of rotation of the top carriage. As a result, large changes in deflection will cause misalinement of the aiming posts because of movement of the sight on an arc around the center of the top carriage. Placing the aiming posts between 800 and 900 mils to the left front (deflection 2400 to 2300) when the howitzer is in center of traverse will minimize the effects of this condition.

42. CORRECTION FOR DISPLACEMENT. When the gunner notes that the vertical hair of the telescope is displaced from the line formed by the two aiming posts (or aiming post lights), he lays in such a manner that the far aiming post (light) appears exactly midway between the near aiming post (light) and the vertical hair (fig. 24). If the displacement is due to traverse of the piece, the gunner continues to lay as described above. However, if the displacement is due to progressive shift of position of the carriage from shock of firing or other cause, the gunner will notify the chief of section, who, at the first lull in firing, will notify the executive and request permission to re-aline aiming posts . To perform the alinement, the piece is laid with the sight picture described above. The far aiming post is moved into alinement with the vertical hair of the telescope, and then the near aiming post is alined. If, due to terrain conditions, it is impracticable to move one of the two aiming posts, the piece is laid for direction and referred to the aiming post which cannot be moved. The other post is alined and new deflection setting reported to the executive.

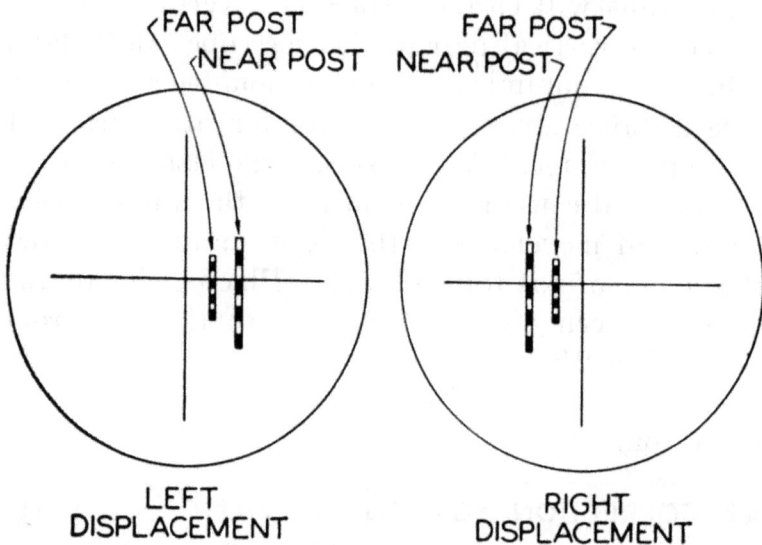

Figure 24. Appearance of aiming posts when piece is laid correctly after displacement.

43. PREPARATION OF POSITION. a. General. (See FM 6–140 and TM 9–331.) To insure stability of the carriage in firing, the piece should be emplaced on level ground, or the position prepared so that when the piece is emplaced the jack float rests on a horizontal plane and neither wheel touches the ground. It is desirable that the areas on which ends of the trails are to be placed be approximately in the same horizontal plane. On soft ground, it will be necessary to support the jack float with a matting of logs, timber, sandbags, or some similar material. In the event that it becomes necessary to fire without emplacing the firing jack, the piece may be fired off the tires without immediate injury to the carriage. If the maximum right traverse is to be obtained when firing off the tires, the firing jack rack plunger must be lowered sufficiently to clear the traversing pinion. However, shortened life of the carriage can be expected if charge 7 is consistently fired at elevations over 530 mils. Firing off the tires is an expedient only.

b. **Digging spade pits.** Unless lack of time prevents, pits will be dug for the spades. On soft ground the spades can be seated by firing, but the displacement of the carriage will be greater than if the spades had been dug in originally.

c. **Digging recoil pits.** When high-angle fire is contemplated, a full 46-inch clearance between the breech and the ground must be provided. This will usually necessitate the digging of a recoil pit. After the piece has been emplaced and, if possible, seated, the piece will be elevated to maximum elevation, and a recoil pit dug that will insure proper clearance throughout the limits of traverse. When such a pit is dug, flooring should be provided over the pit to facilitate the service of the piece.

44. REPORTING ERRORS. Each member of the howitzer section should be constantly impressed with the importance of reporting promptly to the chief of section any errors made by the members of the howitzer section. The chief of section will report errors immediately to the executive, as prescribed in paragraph 19**b**(8).

45. CEASE FIRING. The command CEASE FIRING normally is given to the howitzer squad by the chief of section, but in emergencies anyone present may give the command. At this command, regardless of its source, firing will cease immediately. If the piece is loaded, the chief of section will report that fact to the executive. Firing is resumed at the announcement of the elevation.

46. SUSPEND FIRE. The command SUSPEND FIRE is given when it becomes necessary to interrupt firing prior to the end of a mission. It indicates that adjustment of fire will continue after a short delay. If the piece is loaded, the chief of section will report that fact to the executive. Firing is resumed at the announcement of elevation.

47. CHANGES IN DATA DURING FIRING. Except in continuous fire, the announcement to the howitzer section of any new element of firing data serves as a signal to stop all firing previously ordered but not yet executed. If the piece is not loaded at the announcement of a new element of firing data, the new data will be set off, and firing resumed at the announcement of elevation. If no change in fuze setting is required, or if the piece is loaded with percussion-fuzed shell, the new data are set off, and the firing is resumed. If the pieces are loaded with time-fuzed shell, and the data require a change in fuze setting, that fact will be reported to the executive. In continuous fire, the changes in data are so applied as not to stop the fire or break its continuity.

48. TO UNLOAD THE PIECE. No special unloading rammer is provided for unloading service rounds of ammunition. When it is desired to unload the piece, the projectile normally will be fired out of the weapon. However, the cleaning and unloading rammer M7, affixed to the rammer staff, may be used for unloading lightly stuck projectiles. This will be done only under the immediate supervision of an officer. For additional information see TM 9–331.

49. MISFIRES. (See AR 750–10.) **a. Misfire of a primer.** Should the primer fail to fire, no report is heard. Failure may be due to a defective primer or to failure of the firing pin to strike the primer properly. In any case, at least two attempts will be made to fire the primer before it may be removed. If, upon examination, it is found that the primer is not fired, a new primer will be inserted and fire continued. However, before inserting a new primer, the firing mechanism will be examined for worn or broken parts. If, on the other hand, examination shows that the primer has fired, a new primer will not be inserted nor the breech opened, and no person will be permitted to remain near or pass in rear of the breech until at least 60 seconds has elapsed after firing the primer, when the procedure will be as prescribed in **b** below for a misfire of the charge.

b. Misfire of a charge. After the prescribed interval has elapsed the faulty charge may be removed. The faulty charge must be stored separately from other charges. Before removing the faulty charge, the chief of section should note the position and condition of the charge in the chamber, because if the primer has fired, an abnormal condition of the propelling charge is indicated, such as missing igniter, igniter end of charge against projectile, wet igniter, igniter-protector cap not removed, or igniter charge folded over and not accessible to the flash of the primer. If the cause of the misfire is found, report will be made to the executive who will take appropriate action.

50. CARE OF AMMUNITION. (See TM 9–331 and TM 9–1900.) To insure uniform results in firing, to prolong the life of the tube, and to avoid accident, great care must be exercised in the storage and handling of ammunition at the battery. Provisions of TM 9–1900 applicable to field service should be followed carefully. In actual service, the conditions inherent in each situation will largely determine the amount of time, labor, and materials that must be expended to provide the requisite facilities for the proper handling and storage of ammunition. If the position is to be occupied for a few hours only, a paulin spread on the ground may be sufficient; while in a zone defense, elaborate magazines may be necessary. In all situations, resourcefulness and ingenuity in the utilization of facilities and materials available at the position of the pieces will be necessary. The following principles should be observed:

a. Ammunition must be protected from damage, especially to rotating bands. When it is received, it should be sorted into lots and placed in the best available storage. Ammunition data cards should be retained until after all ammunition pertaining thereto is expended. Fuzes must not be stored with other components, and all components should be kept in their containers until their early use is anticipated. Protection should be provided against moisture, dirt, and direct rays of sun, and, as far as practicable, hostile artillery fire and bombing. Protection against weather, dirt, and sun may be obtained by the use of paulins below and above the ammunition, and suitable dunnage below and between the layers. Protection against hostile fire may be obtained by the use of small dispersed stacks, trenches, or dugouts.

b. Care must be exercised to keep sand and dirt out of the fuze cavity of unfuzed ammunition.

c. All safety precautions in handling ammunition must be rigidly enforced. See AR 750–10 and TM 9–1900.

51. FUZE SETTERS. For care and maintenance of fuze setters, see TM 9–331.

52. SECTION DATA BOARD. When positions are occupied for more than a few hours, data boards may be used by each section for recording such items as base deflection, calibration corrections when appropriate, minimum elevation, data for defensive barrages and counter preparations, and other data the need for which may be urgent.

53. SAFETY PRECAUTIONS. In addition to the safety precautions indicated in this manual, see AR 75010, FM 6–140, TM–9–331, and TM 9–1900.

CHAPTER 7

SECTION MAINTENANCE DRILL

---

54. GENERAL a. Inspection and maintenance are essential to insure that the howitzer section is ready to move and to shoot at all times. Section drill on inspection and maintenance will make the work routine, thorough, and fast.

b. The outline of duties in the section maintenance drill assigns duties to all members of the section. When the section is reduced in strength, the chief of section will assign duties to insure that all maintenance steps are completed.

c. The inspections and maintenance operations performed by the howitzer section are outlined in the following paragraphs. Details of all preventive maintenance services are contained in the technical manuals on the particular equipment and in appropriate Department of the Army Lubrication Orders.

55. DUTIES IN INSPECTION BEFORE OPERATION (MARCH). a. Chief of Section. (1) Supervises detailed inspection of section.

(2) Inspects recoil system for oil leakages.

(3) Inspects oil index to insure that a proper reserve of oil is present in recoil system (TM 9–331).

(4) Verifies ammunition for condition, lot number, and loading.

(5) Verifies loading and security of section equipment.

(6) Verifies gas and oil supply.

(7) Verifies water and ration supply.

(8) Verifies presence of gun book, trip ticket, Standard Form 26 (Driver's Accident Report), technical manuals, and lubrication orders for piece and prime mover.

(9) Verifies coupling of piece and connection of brake lines.

(10) Verifies release of hand brakes.

(11) Causes air brakes to be tested after loading and coupling.

(12) Verifies security of traveling lock and trail lock.

(13) Receives reports of personnel of his section upon completion of their duties.

(14) Reports to battery executive when section personnel have completed their duties. "Sir, No. (so-and-so) in order," or reports any defects that the section cannot remedy without delay.

b. Gunner. (1) Inspects condition, contents, completeness, and security of section chest. See TM 9–331.

(2) Verifies condition and security of sighting and fire-control equipment, including telescope mount. See TM 9–331.

(3) Places piece in center of traverse and elevates or depresses tube so No. 5 can secure traveling lock (if piece is not previously prepared for travel).

(4) Assists the battery mechanic to service the recoil system, if necessary.

(5) Inspects tires and wheels for damage and loose or missing parts; tests air pressure of tires and corrects it if necessary.

(6) Assists Nos. 1 and 2 to remove and replace rear section of howitzer cover.

(7) Reports, "Gunner ready."

c. No. 1. (1) Inspects breechblock, firing mechanism, powder chamber, and bore for cleanliness, freedom from foreign matter, and lubrication.

(2) Inspects brake lines and connections for serviceability and security.

(3) Assists the gunner and No. 2 to remove and replace rear section of howitzer cover.

(4) Inspects security of trail lock and verifies that the lunette is secure in pintle and that pintle latch is closed and locked.

(5) Reports, "No. 1 ready."

d. No. 2. (1) Assisted by the gunner and No. 1, removes and replaces rear section of howitzer cover; inspects fastenings and general condition of cover.

(2) Tests adjustment of hand brakes; inspects after coupling to see that brakes are released.

(3) Assists gunner to test tire pressure and assists in correcting it if necessary.

(4) Reports, "No. 2 ready."

e. No. 3. (1) Verifies presence and security of spade, handspike, and rammer staff sections on right trail.

(2) Assists ammunition corporal in verifying presence, condition, and proper loading of ammunition in section.

(3) Reports, "No. 3 ready."

f. No. 4. (1) Verifies presence and security of spade, handspike, spade keys, and rammer staff sections on left trail, together with the firing-jack float and the loading tray.

(2) Inspects the firing jack and the traveling lock for presence, security, and condition of all parts and appurtenances.

(3) Assists Nos. 5 and 6 to remove and replace front section of howitzer cover.

(4) Reports, "No. 4 ready."

g. No. 5. (1) Verifies presence and security of jack handles on the right shield.

(2) Verifies condition and security of sponge and bucket, and presence of sufficient supply of cleaning and preserving materials.

(3) Assists Nos. 4 and 6 to remove and replace howitzer front cover section.

(4) Reports, "No. 5 ready."

h. No. 6. (1) Assists ammunition corporal in verifying presence, condition, and proper loading of ammunition in section.

(2) Assisted by Nos. 4 and 5, removes and replaces front section of howitzer cover; inspects fastenings and general condition of cover.

(3) Verifies presence and completeness of aiming posts and aiming post lights.

(4) Reports, "No. 6 ready."

i. Nos. 7 and 8. Assist driver of prime mover in inspection and preventive maintenance when directed to do so by the chief of section.

j. Ammunition corporal. (1) Inspects ammunition. Verifies presence of all components and each component for proper type (as directed by the battery executive). Verifies proper loading.

(2) Reports, "Ammunition corporal ready."

k. Driver. (1) Performs "before operation" duties as prescribed for his vehicle.

(2) Reports, "Driver ready."

56. DUTIES IN INSPECTION DURING THE MARCH. a. Chief of section. (1) Rides with driver and supervises march discipline.

(2) Watches vehicle instruments for proper functioning, and listens for abnormal or unusual noises.

(3) Assigned duties for antiaircraft and anti-mechanized security.

b. Gunner. (1) Listens for abnormal or unusual noises.

(2) Watches prime mover instruments and controls for proper functioning when a tractor is used as the prime mover.

(3) Observes towed load for security when a truck is used as the prime mover. Signals chief of section in case of malfunction.

c. Nos. 1 to 8 inclusive. (1) Perform duties as antiaircraft and antimechanized security sentries as assigned by chief of section.

(2) Listen for abnormal or unusual noises in prime mover and howitzer.

(3) Nos. 6 and 7 observe towed load for security when a tractor is used as the prime mover. Signal chief of section in case of malfunction.

d. Driver. Performs "during operation" duties as prescribed for his vehicle.

57. DUTIES IN INSPECTION DURING THE HALT. a. Chief of section. (1) Supervises inspection and maintenance at halt.

(2) Assigns duties of security sentries as necessary.

(3) Insures that personnel remain to the right of the left wheel line except to inspect left wheels and tracks.

(4) Receives reports of personnel upon completion of their duties.

(5) Reports to the battery executive when inspection is completed, "Sir, No. (so-and-so) in order," or reports any defects that the section cannot remedy without delay.

b. Gunner. (1) Verifies presence, condition, and security of sighting equipment, howitzer covers, staff sections, aiming posts, trail handspikes, jack float, loading tray, spades, and spade keys.

(2) Reports, "Gunner ready."

c. No. 1. (1) Inspects coupling of howitzer to prime mover.

(2) Inspects trail traveling lock.

(3) Inspects brake lines and connections.

(4) Inspects connections, functioning, and mounting of stop and tail light.

(5) Reports, "No. 1 ready."

d. No. 2. (1) Inspects tires for wear, bruises, cuts, stones in treads, and for low air pressure.

(2) Inspects wheels for loose or missing nuts, hubcap screws, and valve caps.

(3) Inspects for overheated wheel bearings and brake rums.

(4) Verifies release of hand brakes.

(5) Reports, "No. 2 ready."

e. Nos. 3 to 8 inclusive. Perform duties as assigned by chief of section.

f. Ammunition corporal. (1) Inspects security of ammunition.

(2) Reports, "Ammunition corporal ready."

g. Driver. (1) Performs "at halt" duties as prescribed for his vehicle.

(2) Reports, "Driver ready."

58. DUTIES IN INSPECTION PRIOR TO AND DURING FIRING. a. Chief of section. (1) Supervises and commands section as prescribed in chapter 5.

(2) Supervises servicing of recoil system and test and adjustment of sighting and fire-control equipment prior to firing.

b. Gunner. (1) Tests and adjusts sighting and fire-control equipment prior to firing, provided sufficient time is available.

(2) Performs duties during firing as prescribed in chapter 5.

c. Nos. 1 to 8 inclusive. Perform normal duties incident to firing as prescribed in chapter 5 and any other duties as directed by the chief of section.

d. Ammunition corporal. Performs duties prescribed in chapter 5, assisted by cannoneers.

e. Driver. The driver will move his vehicle to the motor park under the direction of the first sergeant and will perform "after operation" duties prescribed for his vehicle.

59. DUTIES IN INSPECTION AND MAINTENANCE AFTER FIRING. a. Chief of section. (1) Supervises detailed inspection and maintenance of howitzer.

(2) Inspects recoil system for undue leakage and supervises the establishment of the correct oil reserve (TM 9—331).

(3) Inspects all ammunition.

(4) Verifies tools, accessories, and equipment for completeness and condition.

(5) Verifies presence of and current entries in gun book, trip ticket, etc.

(6) Verifies resupply of emergency rations, gasoline, oil, and water.

(7) Receives reports from members of the section as they complete inspection and maintenance operations.

(8) Reports to battery executive, "Sir, No. (so-and-so) in order."

b. Gunner. (1) Verifies presence of and cleans,

oils, and secures all sighting and fire-control equipment.

(2) Verifies contents of section chest and inspects for condition of chest and contents.

(3) Assists Nos. 1 and 2 to remove and replace rear section of howitzer cover.

(4) Inspects bore and breech mechanism after cleaning.

(5) Inspects tires and verifies air pressure, assisted by No. 2.

(6) Reports, "Gunner ready."

c. No. 1. (1) Assists the gunner and No. 2 to remove and replace rear section of howitzer cover.

(2) Removes, cleans, oils, and replaces breech mechanism; receives assistance as directed by the chief of section.

(3) Inspects brake lines and connections for condition and security.

(4) Reports, "No. 1 ready."

d. No. 2. (1) Assists the gunner and No. 1 to remove and replace rear section of howitzer cover; inspects fastenings and general condition of cover.

(2) Inspects hand brakes and wheels for damage and loose or missing parts.

(3) Assists Nos. 3, 4, 5, and 6 to clean howitzer and carriage.

(4) Assists the gunner in inspecting tires and verifying air pressure.

(5) Reports, "No. 2 ready."

e. No. 3. (1) Verifies presence and security of spade, handspike, and rammer staff sections on right trail; insures that these are clean.

(2) Assists Nos. 2, 4, 5, and 6 to clean howitzer and carriage.

(3) Using War Department Lubrication Order, assists No. 4 to lubricate howitzer and carriage.

(4) Inspects carriage for loose or missing nuts, bolts, rivets, and broken welds; looks for excess grease or oil under carriage.

(5) Cleans, lubricates, and returns fuze setter to section chest.

(6) Reports, "No. 3 ready."

f. No. 4. (1) Verifies presence and security of spade, handspike, spade keys, and rammer staff sections on left trail, together with the firing-jack float and the loading tray; insures that these are clean.

(2) Assists Nos. 5 and 6 to remove and replace front sections of howitzer cover.

(3) Assists Nos. 2, 3, 5, and 6 to clean howitzer and carriage.

(4) Using War Department Lubrication Order, and assisted by No. 3, lubricates howitzer and carriage.

(5) Inspects the firing jack and the traveling lock; insures that these are clean.

(6) Reports, "No. 4 ready."

g. No. 5. (1) Verifies presence and security of jack handles on the right shield.

(2) Assists Nos. 4 and 6 to remove and replace front section of howitzer cover.

(3) Procures sponge, bucket, and cleaning and preserving materials and assists Nos. 2, 3, 4, and 6 to clean howitzer and carriage.

(4) Reports, "No. 5 ready."

h. No. 6. (1) Verifies presence and condition of aiming posts and aiming post lights.

(2) Assisted by Nos. 4 and 5, removes and replaces front section of howitzer cover; inspects fastenings and general condition of cover.

(3) Assists ammunition corporal to inspect, clean, and care for ammunition.

(4) Reports, "No. 6 ready."

i. Nos. 7 and 8. Assist driver of prime mover to perform "after operation" inspection and preventive maintenance.

j. Ammunition corporal. (1) Inspects all ammution components.

(2) Supervises and inspects storage of ammunition; accomplishes turn back when applicable.

(3) Reports, "Ammunition corporal ready."

k. Driver. (1) Performs "after operation" duties as prescribed for his vehicle.

(2) Reports, "Driver ready."

60. DUTIES IN WEEKLY INSPECTION AND MAINTENANCE. a. Chief of section. Supervises section in weekly inspection and maintenance services on howitzer, tools, accessories, and equipment. (See TM 9–331 and WDLO.) Obtains services of the artillery mechanic for operations requiring skill and tools beyond the capabilities of the section.

b. Gunner, ammunition corporal, and cannoneers Nos. 1 to 6 inclusive. Perform normal care and cleaning as directed by the chief of section.

c. Driver and Nos. 7 and 8. Assist the motor mechanic in performance of weekly inspection and maintenance services as prescribed for the vehicle.

CHAPTER 8

CARE, MAINTENANCE, AND ADJUSTMENT OF MATÉRIEL

Section I. HOWITZER AND CARRIAGE

61. DUTIES OF PERSONNEL. The duties to be performed by personnel of the howitzer section when maintenance services are being accomplished on the piece are prescribed in chapter 7.

62. INSPECTIONS. Regular inspections are required to insure that matériel is maintained in serviceable condition.

a. The chief of section is responsible for his piece and should make a thorough daily inspection. If he sees the need for a repair or adjustment, he notifies the battery executive immediately.

b. The executive accompanied by the artillery mechanic, makes a daily spot-check inspection. He inspects different parts each day to insure complete coverage of the weapons within a few days. At least once a month, the executive makes a thorough mechanical inspection of the matériel, tools, equipment, and spare parts.

c. Battery, battalion, and higher commanders will make frequent command inspections to assure themselves that the appearance, completeness, and condition of the matériel in their commands are being maintained at prescribed standards.

d. Detailed instructions for inspections of this

weapon are found in TM 9–331. Deficiencies revealed in inspections should be remedied promptly. The artillery mechanic should make authorized repairs or adjustments. Other necessary repairs should be reported to the ordance maintenance company.

63. MAINTENANCE SERVICES. Refer to TM 9–331 for complete details concerning maintenance instructions for the howitzer. Complete lubrication instructions are contained in the pertinent Department of the Army Lubrication Order. Maintenance services pertaining to the prime mover are described in technical manuals pertinent to the equipment.

64. DISASSEMBLY, ADJUSTMENT, AND ASSEMBLY. Authorized disassemblies and adjustments permitted to using personnel are prescribed in TM 9–331, supplemented by instructions contained in ORD 7 SNL C–39, ORD 8 SNL C–39, and ORD 9 SNL C–39. No deviation from these procedures will be permitted except when authorized by the responsible ordnance officer.

65. RECORDS. a. The principal records relative to the matériel are the artillery gun book (O.O. Form 5825), a field report of accidents (AR 750–10), and the unsatisfactory equipment report (WD AGO Form 468). Information regarding the use and purpose of these records is contained in the forms themselves.

b. The chief of section, the battery executive, and the battery commander should maintain records of a semipermanent character for their own information and guidance.

Section II. SIGHTING AND FIRE-CONTROL EQUIPMENT

66. GENERAL. For the description, care, and preservation of the sighting and fire-control equipment issued with the howitzer, see TM 9–331.

67. TEST AND ADJUSTMENT OF SIGHTING AND FIRE-CONTROL EQUIPMENT. For the procedure to be followed in making detailed tests and adjustments of this equipment, refer to appendix II. Tests and adjustments are performed before firing, during lulls in firing, and at other times when it is deemed necessary. Corrections may be carried on scales and bubbles until time is available to make adjustments, provided the corrections fall within allowable limits. Battery personnel are forbidden to disassemble sighting and fire-control equipment.

APPENDIX I

QUALIFICATION OF GUNNERS

---

1. PURPOSE AND SCOPE. This appendix prescribes the procedure to be followed and the tests to be given in the qualification of gunners. The purposes of the tests are—

a. To provide a means of determining the relative proficiency of the individual artillery soldier in the performance of the duties of the gunner, 155-mm Howitzer M1. *The tests will not be a basis for determining the relative proficiency of a battery or higher unit.*

b. To serve as an adjunct to training.

2. GENERAL INSTRUCTIONS. a. Standards of precision. The candidate will be required to perform the tests in accordance with the standards of precision enumerated below.

(1) Scale settings must be exact, and matching indexes must be brought into coincidence.

(2) Level bubbles must be exactly centered.

(3) The vertical hair in the reticle of the panoramic telescope must be alined on the left edge of the aiming post or on exactly the same part of the aiming point or target each time the piece is laid.

(4) Final motions of azimuth and elevation setting knobs, as well as traversing and elevation handwheels, must be made in the appropriate direction as prescribed in TM 9–331 and paragraph 20 of this manual.

b. Assistance. The candidate will receive no unauthorized assistance during the examination. Each candidate may select authorized assistants as indicated in the tests. In the event a candidate fails any test because of the fault of the examiner or any assistant, the test will be disregarded, and the candidate will be given another test of the same nature.

c. Time. The time for any test will be the time from the last word of the command to the last word of the candidate's report. The candidate may begin any test after the first word of the first command.

d. Scoring. Scoring will be conducted in accordance with the two subparagraphs "Penalties" and "Credit" under each subject. If a test is performed correctly, credit will be given in accordance with the subparagraph "Credit" under each subject. The examiner will familiarize himself with the items listed under the subparagraph "Penalties" under each subject.

e. Preparation for the tests. The piece will be prepared for action and the candidate posted at the proper position corresponding to the test being conducted or as indicated in the subparagraphs entitled "Special instructions." The examiner will assure himself that the candidate understands the requirements of each test and will require the candidate to report "I am ready," before each test.

3. OUTLINE OF TESTS.

Paragraph No.	Subject	Number of tests	Points each	Maximum credit
4	Direct laying, panoramic telescope	4	2	8
5	Indirect laying, deflection only	18	2	36
6	Laying for elevation with elevation scale	3	2	6
7	Laying for elevation with gunner's quadrant	3	2	6
8	Displacement correction	2		6
	Part I	(1)	5	(5)
	Part II	(1)	1	(1)
9	Measuring site to the mask	1	4	4
10	Measuring elevation	1	4	4
11	Measuring deflection	1	5	5
12	Test and adjustment of sighting and fire-control equipment	6		10
	Test No. 1	(1)	2	(2)
	Tests Nos. 2, 3, 4, and 5	(4)	1	(4)
	Test No. 6	(1)	4	(4)
13	Matériel	3	5	15
	Total credit			100

4. DIRECT LAYING, PANORAMIC TELESCOPE. a.
Scope of tests. Four tests (two groups of two tests each) will be conducted in which the candidate will be required to execute commands similar to those given below. Tests Nos. 1 and 2 (likewise tests Nos. 3 and 4) will be executed as one series of commands.

b. Special instructions. (1) A stationary target will be placed approximately 600 yards from the piece.

(2) The candidate will be posted as gunner.

(3) An assistant, selected by the candidate, will be posted as No. 3 to set the announced elevation on the elevation scale and to keep the sight mount cross leveled.

(4) The piece will be pointed so that—

(*a*) A shift of approximately 100 mils is required for tests Nos. 1 and 3.

(*b*) The trails need not be shifted for any of the four tests.

(5) Laying at the termination of tests Nos. 1 and 3 will not be disturbed at the beginning of tests Nos. 2 and 4.

(6) The examiner will announce the assumed direction of movement of the target before tests Nos. 1 and 3. The assumed direction of movement of the target in test No. 3 will be opposite to that in test No. 1.

c. Outline of tests.

Test No.	Examiner commands (for example)	Action of candidate
1 and 3	TARGET THAT TANK, LEAD 5, ELEVATION 16.	Matches indexes on rotating head of panoramic telescope. Sets azimuth scale and micrometer of panoramic telescope at zero. Traverses piece until proper lead in mils has been set. Places horizontal reticle line on the base of the target. When No. 3 calls "Ready," commands FIRE and steps clear.
2 and 4	RIGHT (LEFT) 4, ADD (DROP) 7.	Same as test No. 1 above.

d. Penalties.   No credit will be allowed if, after each test—

(1) The indexes on the rotating head of the panoramic telescope are not in coincidence.

(2) The azimuth scale and micrometer of the panoramic telescope are not set at zero.

(3) The lead in mils is not indicated properly.

(4) The horizontal reticle line is not on the base of the target.

e. Credit.

Time in seconds, exactly or less than	6	7	8
Credit	2.0	1.5	1.0

5. INDIRECT LAYING, DEFLECTION ONLY.   a. Scope of tests.   Eighteen tests (two groups of nine tests each) will be conducted in which the candidate will be required to execute commands similar to those given below.   Tests Nos. 1 through 9 (likewise tests Nos. 10 through 18) will be executed as one series of commands.

b. Special instructions.   (1) Commands will not necessitate shifting trails.

(2) The examiner will select a suitable aiming point and identify it to the candidate.

(3) Commands for deflection difference will be given *only* in the tests indicated in the examples below.

(4) The command for deflection change for each test will be within the following prescribed limits:

Test No.	Maximum change (mils)	Minimum change (mils)
2 and 11	180	140
3 and 12	90	70
4 and 13	40	20
6 and 15	260	210
7 and 16	100	60
8 and 17	50	30
9 and 18	20	10

(5) The piece will be laid with the correct settings at the conclusion of each test before proceeding with the next test.

(6) Aiming posts will be set out at prescribed distances for this test.

(7) The examiner will designate the section number of the howitzer to be used. The examiner will announce deflection difference commands which will require the application of deflection difference by the candidate.

c. Outline of tests.

Test No.	Examiner commands (for example)	Action of candidate
1 and 10	AIMING POINT, CHURCH STEEPLE TO LEFT FRONT, DEFLECTION 2890, ON NO. 2 OPEN 6.	Sets deflection and applies deflection difference appropriate for his piece. Centers cross-level and longitudinal-level bubbles. Traverses piece until vertical hair is on the aiming point. Verifies centering of bubbles. Re-lays if necessary. Calls "Ready" and steps clear.
2 and 11	RIGHT 180 - - - - - - - - - - - - -	Sets deflection change. Lays on aiming point. Verifies centering of bubbles. Re-lays if necessary. Calls "Ready" and steps clear.
3 and 12	LEFT 70	Same as test No. 2 above.
4 and 13	RIGHT 25, ON NO. 4, CLOSE 3	Same as test No. 2 above, but also applies deflection difference for his piece.
5 and 14	AIMING POINT, AIMING POSTS, REFER, RECORD BASE DEFLECTION.	Refers telescope to aiming posts. Reads base deflection and calls "Base deflection No. (so-and-so), (so much)." Records base deflection on shield and steps clear.

Test No.	Examiner commands (for example)	Action of candidate
6 and 15----	BASE DEFLECTION LEFT 240, ON NO. 4, OPEN 7.	Same as test No. 4 above.
7 and 16----	RIGHT 80------------------	Same as test No. 2 above.
8 and 17----	LEFT 40-------------------	Same as test No. 2 above.
9 and 18----	RIGHT 18, ON NO. 2, CLOSE 4-	Same as test No. 4 above.

d. Penalties. (1) No credit will be allowed if, after each test—

(*a*) The deflection is set incorrectly.

(*b*) The cross-level or longitudinal-level bubble is not centered.

(*c*) The vertical hair of telescope is not on the aiming point or aiming posts, as the case may be, after the cross-level and longitudinal-level bubbles are centered.

(2) No credit will be allowed if the last motion of traverse was not made to the right.

e. Credit. Time in seconds, exactly for less than—

Tests Nos. 1, 10, 6, and 15, each_____	13	14	15
Other tests, each_____	10	11	12
Credit_____	2. 0	1. 5	1. 0

6. LAYING FOR ELEVATION WITH ELEVATION SCALE.

a. Scope of tests. Three tests will be conducted in which the candidate will be required to execute commands similar to those given below.

b. Special instructions. (1) The elevation setting on the elevation scale prior to test No. 1 will be within 40 mils of the initial elevation for the test.

(2) Each test will require a change of settings and the accompanying laying of the tube in elevation within the limits of 20 to 40 mils.

(3) Commands for elevation for tests Nos. 2 and 3 will not be made in the multiples of 5 mils.

c. Outline of tests.

Test No.	Examiner commands (for example)	Action of candidate
1	ELEVATION 420____	Sets announced elevation. Centers cross-level and longitudinal-level bubbles. Calls "Ready" and steps clear.
2	ELEVATION 446____	Same as test No. 1 above.
3	ELEVATION 479____	Same as test No. 1 above.

d. Penalties. (1) No credit will be allowed if, after each test—

(*a*) The elevation is set incorrectly.

(*b*) The cross-level or longitudinal-level bubble is not centered.

(2) No credit will be allowed if the last movement of the elevating handwheel was not made in the direction which increased the elevation of the tube.

e. Credit.

Time in seconds, exactly or less than__	4	5⅗	6¼
Credit_____	2. 0	1. 5	1. 0

7. LAYING FOR ELEVATION WITH GUNNER'S QUADRANT.

a. Scope of tests. Three tests will be conducted in which the candidate will be required to execute commands similar to those given below.

b. Special instructions. (1) The gunner's quadrant will be set at zero for the first test.

(2) Each succeeding test will require a change of quadrant elevation setting within the limits of 30 to 60 mils.

(3) The candidate will be posted to the right of and facing the breech, with the gunner's quadrant in his hand.

(4) An assistant, selected by the candidate, will be posted to the left of the breech to operate the elevating handwheel.

c. Outline of tests.

Test No.	Examiner commands (for example)	Action of candidate
1	QUADRANT 180____	Sets quadrant elevation on gunner's quadrant. Seats quadrant. Has assistant elevate or depress the tube until the quadrant bubble is centered. Calls "Ready" and waits for examiner to verify laying.
2	QUADRANT 240____	Same as test No. 1 above.
3	QUADRANT 207____	Same as test No. 1 above.

d. Penalties. (1) No credit will be allowed if, after each test—

(*a*) The quadrant elevation is set incorrectly.

(*b*) The quadrant is not properly seated.

(*c*) The quadrant bubble is not properly centered.

(2) No credit will be allowed if the last movement of the elevating handwheel was not made in the direction which increased the elevation of the tube.

e. Credit.

Time in seconds, exactly or less than___	8	8⅗	9
Credit_____	2.0	1.5	1.0

8. DISPLACEMENT CORRECTION. a. Scope of test.
One test consisting of two parts will be conducted in which the candidate will be required to execute the commands given below.

b. Special instructions. (1) Aiming posts will be set out at prescribed distances.

(2) An assistant, selected by the candidate, will be stationed close to the far aiming post.

(3) The examiner will require the candidate to lay the piece on an announced deflection and report "I am ready."

(4) The far post or the piece is then moved so that a displacement of 5 to 10 mils occurs.

(5) The laying of the piece at the termination of part I is not disturbed for part II.

c. Outline of test. (1) *Part I.*

Examiner commands	Action of candidate
CORRECT FOR DISPLACEMENT.	Lays the piece so that the far post appears midway between the near post and vertical cross hair of the telescope.
	Verifies centering of bubbles.
	Re-lays if necessary.
	Calls "Ready" and steps clear.

(2) *Part II.*

Examiner commands	Action of candidate
RECORD BASE DEFLECTION, ALINE AIMING POSTS.	Records base deflection on shield and announces "Base deflection (so much), recorded."
	Directs assistant in alining aiming posts.
	Calls "Ready" and steps clear.

d. Penalties. No credit will be allowed if—

(1) *Part I.* (*a*) The far aiming post does not appear midway between near post and vertical cross hair.

(*b*) The cross-level or longitudinal-level bubble is not centered.

(*c*) The last motion of traverse was not made to the right.

(2) *Part II.* (*a*) The recorded base deflection is not the one set on the azimuth scales of the telescope.

(*b*) The aiming posts are not properly alined.

(*c*) The vertical hair of the telescope is not on the aiming posts.

e. Credit.

Part I, time in seconds, exactly or less
  than_____ 6   7   8   9
Credit_____ 5.0 4.0 3.0 2.0
Part II, no time limit.
Credit_____ 1.0 ___ ___ ___

9. MEASURING SITE TO THE MASK. a. Scope of test.
One test will be conducted in which the candidate
will be required to execute the command given below.

b. Special instructions. (1) The piece, prepared
for action, will be placed 200 to 400 yards from a
mask of reasonable height.

(2) The tube will be elevated so that it is 100 to
150 mils above the crest.

(3) An assistant, selected by the candidate, will
be stationed at the post of the gunner to operate the
elevating mechanism.

c. Outline of test.

Examiner commands	*Action of candidate*
MEASURE SITE TO MASK.	Sights along lowest element of bore and has his assistant operate elevating mechanism until line of sighting just clears crest.
	Centers the longitudinal-level bubble by turning the elevation knob and centers the cross-level bubble.
	Reads elevation from elevation scale and micrometer.
	Reports "Site to mask, No. (so-and-so), (so much)."

d. Penalties. No credit will be allowed if—

(1) The line of sighting along the lowest element
of the bore does not just clear crest.

(2) The cross-level or longitudinal-level bubble
is not properly centered.

(3) The site is announced incorrectly.

(4) The last movement of the tube was not made in the direction of increasing elevation.

e. Credit.

Time in seconds, exactly or less than	14	15	16	17
Credit	4.0	3.0	2.0	1.5

10. MEASURING ELEVATION. a. Scope of test.
One test will be conducted in which the candidate will be required to measure the elevation by means of the gunner's quadrant.

b. Special instructions. Prior to the test the examiner will lay the tube at a selected elevation, measure the elevation, and then set the gunner's quadrant at zero.

c. Outline of test.

Examiner commands	Action of candidate
MEASURE THE ELEVATION.	Places gunner's quadrant on quadrant seats of the breech ring.
	Levels bubble by raising or lowering the index arm and turning the micrometer knob.
	Announces "Elevation No. (so-and-so), (so much)," and hands quadrant to examiner.

d. Penalties. No credit will be allowed if—

(1) The quadrant bubble is not centered when the quadrant is seated properly.

(2) The elevation is announced incorrectly.

e. Credit.

Time in seconds, exactly or less than	8	9⅖	10⅗
Credit	4.0	3.0	2.0

11. MEASURING DEFLECTION. a. Scope of test.
One test will be conducted in which the candidate will

be required to measure and report a deflection in accordance with the command given below.

b. Special instructions. (1) The piece will be laid on aiming posts to the left front.

(2) An aiming point within 200 mils to the left or right of the aiming posts will be designated by the examiner and identified by the candidate.

c. Outline of test.

Examiner commands	*Action of candidate*
NUMBER (SO-AND-SO), AIMING POINT, THAT (SO - AND - SO), MEASURE THE DEFLECTION.	Centers cross-level and longitudinal-level bubbles.
	Refers to aiming point.
	Verifies centering of bubbles and relays telescope if necessary.
	Reads deflection and reports "Deflection No. (so-and-so), (so much)," and steps clear.

d. Penalties. No credit will be allowed if—

(1) The cross-level or longitudinal-level bubble is not properly centered.

(2) The vertical hair of the telescope is not on the aiming point.

(3) The deflection is announced incorrectly.

(4) The traversing handwheel is turned.

e. Credit.

Time in seconds, exactly or less than	6	7⅗	8	8⅘
Credit	5.0	4.0	3.0	2.5

12. TEST AND ADJUSTMENT OF SIGHTING AND FIRE-CONTROL EQUIPMENT.

a. Scope of tests. Six tests will be conducted in which the candidate will be required to demonstrate the methods employed in making the prescribed tests and authorized adjustments, or describe the action taken (that is, send to the ordnance maintenance company) if adjustment is not authorized to be made by using personnel.

b. Special instructions. (1) The piece will be prepared for action, the trunnions leveled, and the tube placed in the center of traverse. The muzzle and breech bore sights will be installed in the tube.

(2) For equipment required for making the tests and adjustments see appendix II.

(3) The candidate will select an assistant who will operate the elevating handwheel at the direction of the candidate during tests Nos. 1 and 2, and adjust and aline the testing target at the direction of the candidate prior to test No. 6.

(4) The tests will be conducted in the chronological sequence indicated in **c** below. After the completion of test No. 2, the gunner's quadrant used in tests Nos. 1 and 2 will be used for tests Nos. 3 and 4, with the proper correction, as determined in test No. 1, carried on the quadrant, provided the correction does not exceed 0.4 mil.

(5) Adjustments which the candidate may be required to accomplish will fall within the following limits:

(*a*) Elevation scale, telescope mount, not to exceed one 100-mil graduation.

(*b*) Elevation micrometer scale, telescope mount, not to exceed ten 1-mil graduations.

(*c*) Rotating head elevation indexes, panoramic sight, no adjustment permitted.

(*d*) Rotating head elevation micrometer indexes, panoramic sight, not to exceed one-fourth turn.

(*e*) Panoramic telescope azimuth scale, not to exceed one 100-mil graduation.

(*f*) Panoramic telescope azimuth micrometer scale, not to exceed ten 1-mil graduations.

(*g*) Actuating arm pivot, not to exceed 10 mils out of alinement with bore.

(6) The tube will be leveled at the conclusion of test No. 3 and will not be disturbed thereafter.

c. Outline of tests.

Test No.	Examiner commands	Action of candidate
1	PERFORM END-FOR-END TEST ON GUNNER'S QUADRANT.	Performs test as prescribed in paragraph 4a, appendix II. Calls "Error (so many) mils, quadrant serviceable (unserviceable)" and hands quadrant to examiner for verification.
2	PERFORM MICROMETER TEST ON GUNNER'S QUADRANT.	Performs test as prescribed in paragraph 4b, appendix II. Calls "Quadrant micrometer is (is not) in error."
3	TEST ACTUATING ARM PIVOT------	Performs test and makes adjustment, if necessary, as prescribed in paragraph 5, appendix II. Calls "Ready" when adjustment is complete.
4	TEST LEVELS ON TELESCOPE MOUNT.	Performs test as prescribed in paragraph 6b, appendix II. Calls "Longitudinal-(cross-) level bubble (s) within (without) allowable limit.

Note. The allowable limit is exceeded if the bubbles are not centered by over one graduation on the level vial. This error should be noted and carried as a correction until ordnance personnel can make the adjustment (fig. 31).

Caution: Do not turn cross-leveling or elevation knobs after this test.

5 TEST THE ELE-VATION SCALES.	Performs test and makes adjustments, if necessary, as prescribed in paragraph 7, appendix II. Calls "Ready" when adjustments are complete.

Note. Prior to test No. 6, the cross-leveling and longitudinal-leveling of the tube and the panoramic telescope mount will be verified by the examiner, and the testing target will be alined by the candidate with the help of his selected assistant as prescribed in paragraph 8a, appendix II.

6 BORESIGHT THE HOWITZER.	Performs tests and makes adjustments as prescribed in appendix II, paragraph 8. (The deflection adjustment may be made by either method as prescribed by the examiner.) Calls "Ready" when adjustments are complete.

d. Penalties. (1) *General.* The tests are not essentially speed tests. The purpose of the prescribed time limits is to insure that the candidate can perform the operation without wasted effort.

(2) *Test No. 1.* No credit will be allowed if—

(*a*) The bubble of the gunner's quadrant does not center when verified by the examiner.

(*b*) The error (one-half the amount of the angle which was indicated when the quadrant was first reversed and the bubble centered by moving the index arm and micrometer) is incorrectly given by the candidate.

(*c*) The candidate fails to declare the quadrant unserviceable if the error (necessary correction) exceeds 0.4 mil.

(*d*) The time to complete the test exceeds 2 minutes.

(3) *Test No. 2.* No credit will be allowed if—

(*a*) The procedure was not followed correctly.

(*b*) The time to complete the test exceeds 1 minute.

(4) *Test No. 3.* No credit will be allowed if—

(*a*) The bubble of the gunner's quadrant is not centered.

(*b*) The candidate does not call correctly in regard to the status of either the cross-level or longitudinal-level bubbles.

(*c*) The time to complete the test exceeds 1 minute.

(5) *Test No. 4.* No credit will be allowed if—

(*a*) The procedure was not followed correctly.

(*b*) The time to complete the test and adjustments exceeds 3 minutes.

(6) *Test No. 5.* No credit will be allowed if—

(*a*) The zero line of either the elevation scale or the elevation scale micrometer is not in coincidence with its respective index.

(*b*) The time to complete the test and adjustments exceeds 2 minutes.

(7) *Test No. 6.* No credit will be allowed if—

(*a*) The candidate fails to make any adjustment when such adjustment is indicated.

(*b*) The rotating head elevation micrometer indexes are not in coincidence.

(*c*) The zero line of either the azimuth scale or the azimuth scale micrometer is not in coincidence with its respective index.

(*d*) The center line of the bore, as viewed through the bore sights, or the line of sight of the telescope do not fall on their respective sighting points on the testing target when all scales are set at zero.

(*e*) The time to complete the tests and adjustments exceeds 4 minutes and 30 seconds.

e. Credit. (1) The candidate will be scored on the general merit of his work in addition to the specific requirements above.

(2) If the tests and adjustments are performed correctly within the prescribed time limit, maximum credit will be given as follows:

Test No. 1 _____ 2
Test No. 2 _____ 1
Test No. 3 _____ 1
Test No. 4 _____ 1
Test No. 5 _____ 1
Test No. 6 _____ 4

Total _____ 10

13. MATÉRIEL. a. Scope of tests. Three tests will be conducted as outlined below which the candidate will be required to perform.

b. Special instructions. (1) For tests Nos. 1 and 2, a paulin will be placed on the ground convenient for the use of the candidate in laying out the disassembled parts. The candidate will be allowed to select the tools and accessories necessary for the performance of the tests prior to the start of the tests.

(2) The candidate may select an assistant to aid him in lowering and lifting the breechblock.

(3) For test No. 3, a complete set of lubrication equipment authorized for use by battery personnel will be made conveniently available on a paulin adjacent to the howitzer. Every type of lubricant used on the howitzer will be placed conveniently available in plainly labeled containers.

c. Outline of tests.

Test No.	Examiner commands	Action of candidate
1	DISASSEMBLE BREECH AND FIRING MECHANISMS.	Performs the operation as described in TM 9–331, laying the parts on the paulin. After disassembly, identifies all parts to examiner.
2	ASSEMBLE FIRING AND BREECH MECHANISMS.	Performs the operation as described in TM 9–331.
3	DAILY, WEEKLY, AND MONTHLY LUBRICATION TEST.	Selects proper lubricating equipment and lubricant and shows *how* and with *which* *lubricant* each lubrication point is serviced. (Actual lubrication is not performed.)

d. Penalties. (1) The tests are not essentially speed tests. The purpose of the maximum time limits allowed is to insure that the candidate can perform the operation without wasted effort.

(2) No credit will be given if the following time limits are exceeded:

	Minutes
Test No. 1	8
Test No. 2	12
Test No. 3	5

(3) A penalty of one-half point will be assessed for each component part not correctly identified or omitted in test No. 1. There is no time limit imposed on the identification of component parts. However, the examiner may reduce the grade if it becomes obvious that the candidate is not familiar with the nomenclature.

(4) A penalty of one-half point will be assessed for each lubrication point missed or lubricated improperly, and for each time the proper lubricating device or proper lubricant is not selected.

e. Credit. (1) The candidate will be scored on the general merit of his work in addition to the specific requirements above.

(2) If each test is performed correctly within the prescribed time limit, a maximum credit of 5 points will be given for that test.

APPENDIX II

TEST AND ADJUSTMENT OF SIGHTING AND FIRE-CONTROL EQUIPMENT

---

1. PURPOSE AND SCOPE. The purpose of this appendix is to outline the procedures in making the tests and adjustments of the sighting and fire-control equipment issued with the piece. The tests and adjustments given herein are permitted to competent battery personnel and, if performed correctly, will insure accuracy of the equipment in laying the piece for elevation and direction. The procedure should follow the sequence given below.

2. EQUIPMENT FOR TESTING. The following equipment will be required for performing the tests, and making the adjustments if necessary: bore sights, testing target with a suitable means of supporting it at the proper height and in correct alinement, gunner's quadrant, suitable screw drivers and wrenches, steel plate with two parallel surfaces or a suitable piece of glass, and a plumb line. (The last two items are not issued as items of equipment and must be improvised in the field.)

3. GENERAL CONSIDERATIONS. a. The on-carriage sighting equipment is in adjustment when (with the tube and trunnions level, all scales at zero, all indexes in coincidence, and the longitudinal-level and cross-level bubbles of the telescope mount centered or within the allowable error) the line of sighting of the panoramic telescope is parallel to the axis of the bore.

b. In boresighting the piece a distant aiming point (a sharply defined object at least 2500 yards from the piece) may be used in lieu of the issue testing target. In this case the leveling of the piece is not necessary.

c. When leveling the tube longitudinally (along the axis of the bore) the gunner's quadrant is seated on the leveling plates which are inlaid on the top surface of the breech ring. The piece is leveled traversely (at right angles to the axis of the bore) by placing a flat plate of steel or a piece of plate glass on the counterrecoil piston rod lug and seating the gunner's quadrant on the plates (fig. 25). Prior to starting the tests the trunnions should be level

Figure 25. Cross-leveling the piece by means of the gunner's quadrant.

and the piece placed in the center of traverse, except as indicated in **b** above.

4. TEST OF GUNNER'S QUADRANT. a. End-for-end test.

(1) Set both the index arm and micrometer scale at zero, making sure the auxiliary indexes match (fig. 26).

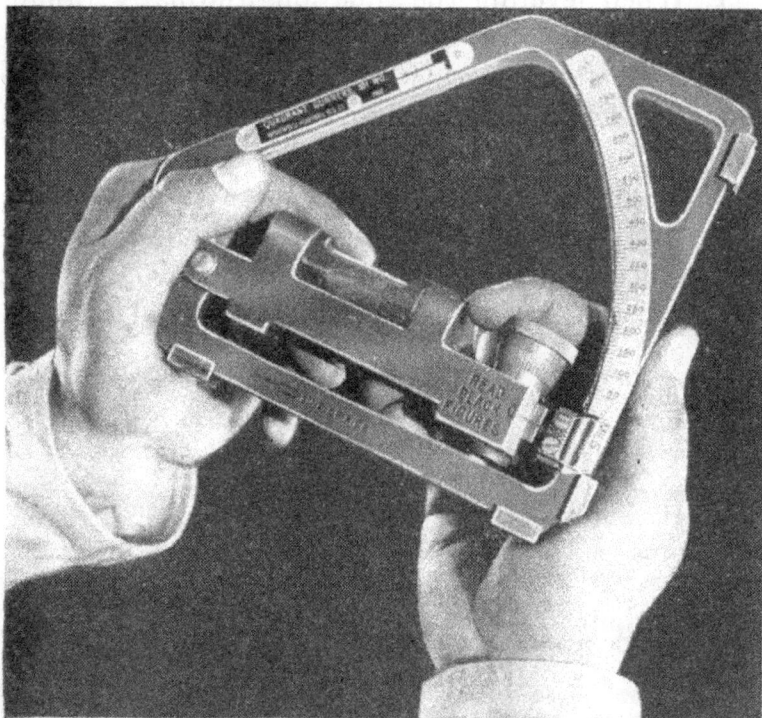

Figure 26. Setting scales of gunner's quadrant at zero.

(2) Place the quadrant on the leveling plates on the breech ring, line-of-fire arrow pointing toward the muzzle, and center the quadrant bubble by manipulating the elevating handwheel (fig. 27).

(3) Reverse the quadrant on the leveling plates (turn it end for end). If the bubble recenters, the quadrant is in adjustment and the test is complete.

Figure 27. Centering level bubble of gunner's quadrant with scales at zero.

(4) If the bubble does not recenter, center it by turning the micrometer knob (setting the index arm to read minus 10 mils if necessary) (fig. 28).

(5) Divide the micrometer reading (subtracting the reading from 10 if the index arm was set to minus 10 mils) by two and set the result on the micrometer scale. This deviation from zero (plus or minus) must be applied to the quadrant to make it a true test level and is called the correction.

(6) With the correction applied to the quadrant, repeat the end-for-end test ((2) and (3) above). The bubble should now center when the quadrant is reversed; if it does not, the complete test must be repeated.

(7) The gunner's quadrant is serviceable if the error (same amount as the correction applied to make the bubble center in both directions) does not exceed

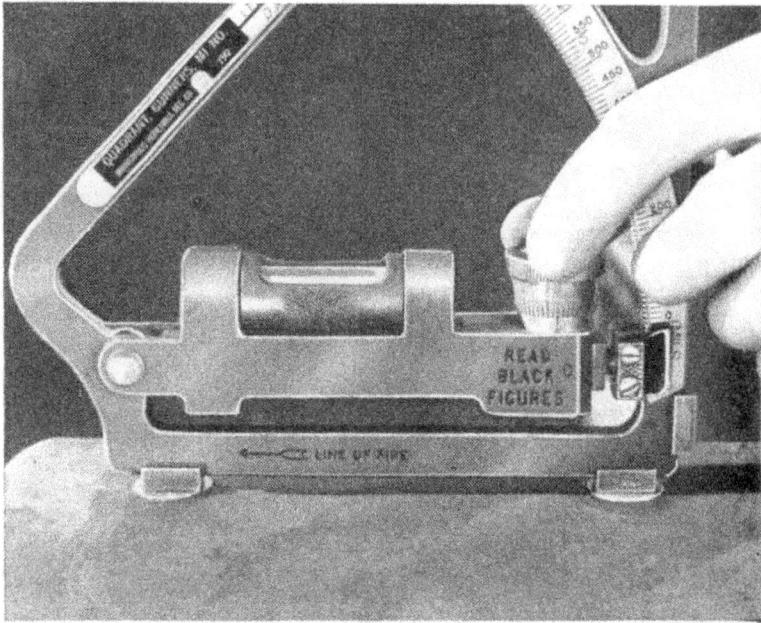

Figure 28. Centering level bubble of gunner's quadrant, quadrant reversed on leveling plates.

0.4 mil. If the allowable error is exceeded it must be adjusted by ordnance personnel.

b. Micrometer test. (1) Set the index arm to read 10 mils on the graduated arc and set the micrometer scale at zero.

(2) Place the quadrant on the leveling plates on the breech ring, line-of-fire arrow pointing toward the muzzle, and center the quadrant bubble by elevating the tube.

(3) Set the index arm at zero on the graduated arc and turn the micrometer one revolution to read 10 mils.

(4) Reseat the quadrant on the leveling plates. The bubble should center.

Caution: Do not disturb the laying of the tube.

(5) If the bubble does not center the micrometer

is in error and must be adjusted by ordnance personnel.

5. TEST AND ADJUSTMENT OF ACTUATING ARM PIVOT. a. Purpose of test.

The purpose of this test and adjustment is to determine if the actuating arm pivot of the telescope mount is parallel to the axis of the bore in elevation, and to make it so if necessary. There is no test permitted to using personnel to determine the relationship in deflection.

b. Outline of test. (1) Place the tube at an arbitrary elevation, preferably not at zero.

(2) Measure the elevation with the gunner's quadrant seated on the leveling plates on the breech ring.

(3) Without disturbing the laying of the piece or the setting on the gunner's quadrant, seat the quadrant on the actuating arm leveling plates (between the telescope mount and the trunnion).

(4) If the quadrant bubble centers, the actuating arm pivot is in adjustment and the test is complete.

(5) If the quadrant bubble does not center, loosen the bracket clamping studs which fasten the actuating arm to the actuating arm bracket on the cradle and center the quadrant bubble by turning the eccentric pin located midway between the clamping studs.

(6) Retighten the clamping studs, being careful to avoid throwing the bubble off center.

(7) Verify the adjustment by repeating the test.

Note. The tube should be level both longitudinally and transversely for the remainder of the tests, except as provided in paragraph 3b above. The purpose of this is to insure the correct offset of the line of sighting of the panoramic telescope from the axis of the bore once the telescope socket is vertical.

6. TEST OF LEVELS ON TELESCOPE MOUNT. a. Purpose of test.   The purpose of this test is to determine if the longitudinal and cross levels on the telescope mount are in adjustment and to erect the telescope socket to a vertical position.

b. Leveling plate method.   (1) set the gunner's quadrant at zero, or at the setting as determined in the end-for-end test in paragraph 4a above.

(2) Place the steel plate (plate glass) on top of the telescope socket and hold the quadrant on top of the plate parallel to the longitudinal-level bubble (fig. 29).

(3) Center the cross-level bubble.

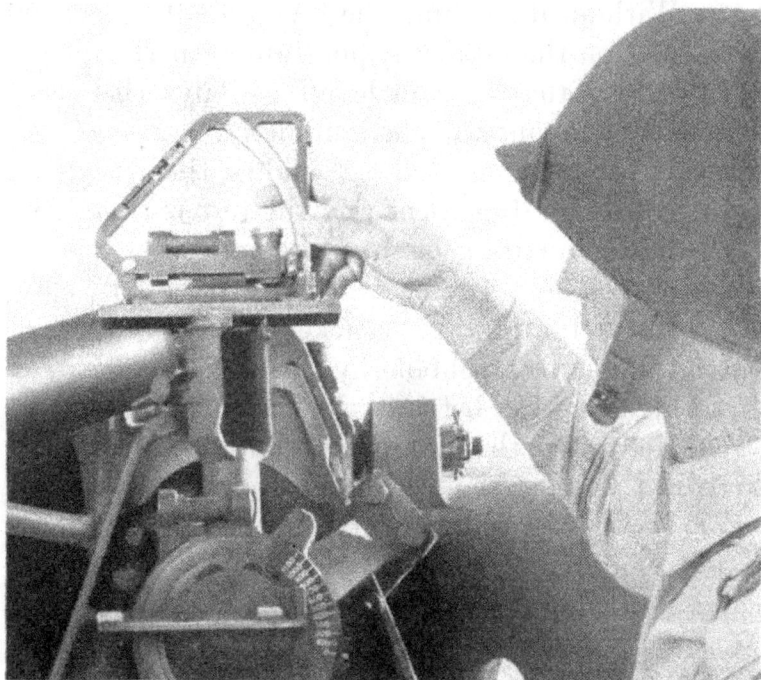

Figure 29. Testing levels on telescope mount, gunner's quadrant on steel plate paralled to longitudinal-level bubble.

(4) Center the bubble of the gunner's quadrant by turning the elevation knob.

(5) Note the position of the longitudinal-level bubble. The allowable limit of error is exceeded if the bubble fails to center by more than one graduation of the level vial. If the bubble is within the allowable limit of error this error may be applied as a correction until ordnance personnel can make the adjustment.

(6) Turn the plate and the gunner's quadrant until they are parallel to the cross-level bubble and center the quadrant bubble by turning the cross-leveling knob.

Caution: Do not disturb the elevation knob. See figure 30.

Figure 30. Testing levels on telescope mount, leveling bubble of gunner's quadrant by turning cross-leveling knob.

(7) Note the position of the cross-level bubble. For allowable limit of error see (5) above (fig. 31).

Caution: Do not turn the cross-leveling or elevation knobs after this test.

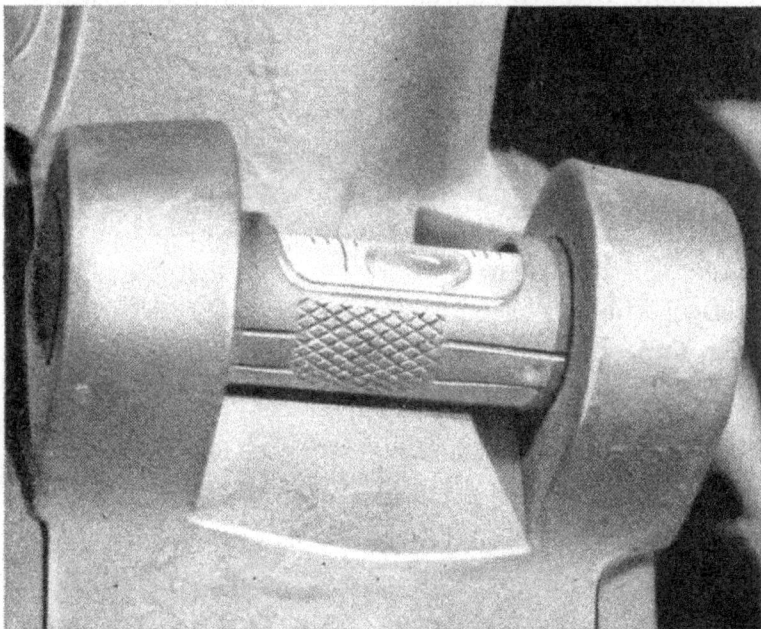

Figure 31. Cross-level bubble of telescope mount, showing error in excess of allowable limit.

c. Plumb line method. (1) Place the panoramic telescope in the telescope socket, set all scales and indexes at zero, and center the cross-level and longitudinal-level bubbles.

(2) Suspend a plumb line so that it coincides with the vertical hair of the panoramic telescope reticle, and is as close to the telescope as practicable.

(3) Elevate and depress the line of sight of the panoramic telescope by turning the rotating head elevation knob throughout its limit of travel.

(4) If the reticle follows the plumb line throughout this range the cross level is in adjustment. If

not, turn the cross-leveling knob until it does follow and note the position of the cross-level bubble. For the allowable limit of error see **b**(5) above.

(5) Turn the rotating head 1600 mils to the left, suspend the plumb line as in (2) above, and repeat the procedure in (3) above.

Caution: Do not disturb the cross-leveling knob.

(6) If the reticle follows the plumb line the longitudinal level is in adjustment. If not, turn the elevation knob on the telescope mount until it does follow and note the position of the longitudinal-level bubble. For the allowable limit of error see **b**(5) above.

Caution: Do not turn the cross-leveling or elevation knobs after this test.

7. TEST AND ADJUSTMENT OF ELEVATION SCALES.

a. At the conclusion of the test in paragraph 6 above, note the readings of the elevation and elevation micrometer scales.

b. If the elevation scale does not read zero (fig. 32), loosen the clamping screws (fig. 33) and adjust the scale to bring the zero graduation into coincidence with its index (fig. 34). Tighten the screws and verify the zero reading.

c. If the elevation micrometer scale does not read zero, loosen the clamping screws in the end of the elevation knob (fig. 35) and adjust the scale to bring the zero graduation into coincidence with its index (fig. 36). Tighten the screws and verify the zero reading.

Caution: Hold the knob firmly while shifting the scale to prevent the displacement of the longitudinal-level bubble.

Figure 32. Elevation scale out of adjustment.

Figure 33. Loosening elevation scale clamping screws.

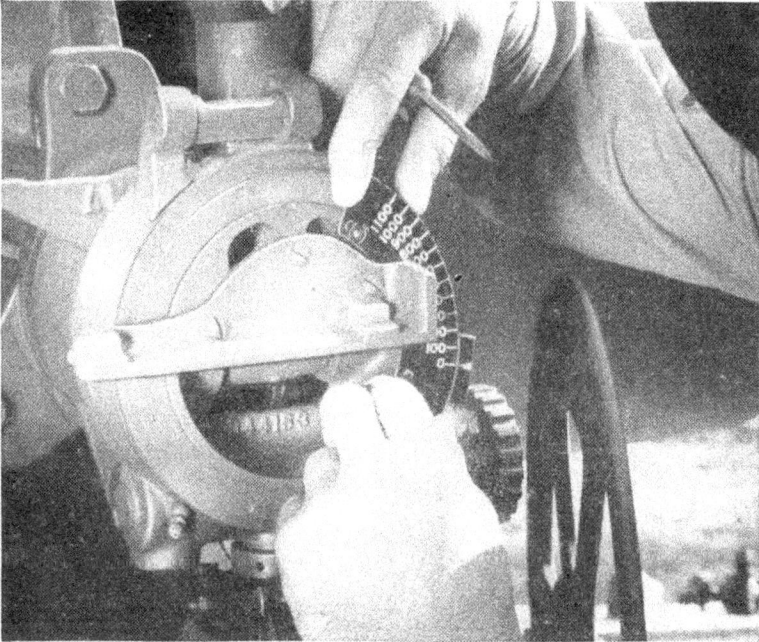

Figure 34. Adjusting elevation scale to read zero.

Figure 35. Loosening elevation micrometer knob clamping screws.

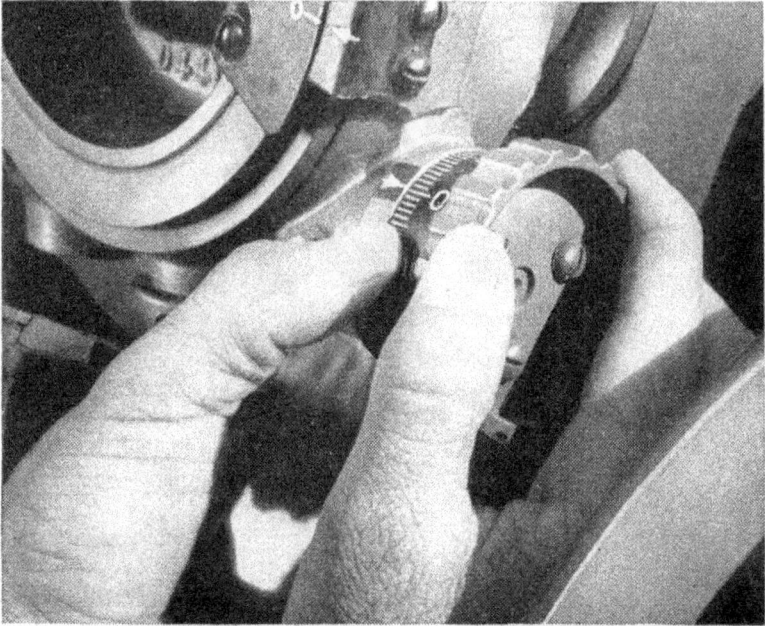

Figure 36. Adjusting elevation micrometer scale to read zero.

8. TEST AND ADJUSTMENT OF PANORAMIC TELE-SCOPE. a. Preliminary operations. (1) The breech and muzzle boresights are placed in their proper positions (or use improvised cross hairs on the muzzle and sight through the primer vent of the closed breechblock).

(2) Place the testing target about 50 yards in front of the muzzle and aline the testing target so that the line of sight through the bore falls on the proper point on the testing target (fig. 37). It is essential that the surface of the testing target is perpendicular to the line of sight and that the vertical lines of the testing target diagram are truly vertical. See TM 9–331.

Figure 37. Alining the testing target for boresighting the howitzer.

(3) The panoramic telescope is placed in the socket and tested for a firm sliding fit (fig. 38). If the telescope is loose it may be tightened when the adjustment described in **c**(1) (*c*) below is made (fig. 39). Verify the position of the telescope mount bubbles.

(4) **If** a distant aiming point is used as the testing target, the line of sight through the bore is alined on the target by manipulating the elevating and traversing handwheels. *The elevation and cross-leveling knobs of the telescope mount are not disturbed after the test prescribed in paragraph 6 above.*

Figure 38. Placing panoramic telescope in telescope mount, testing for fit.

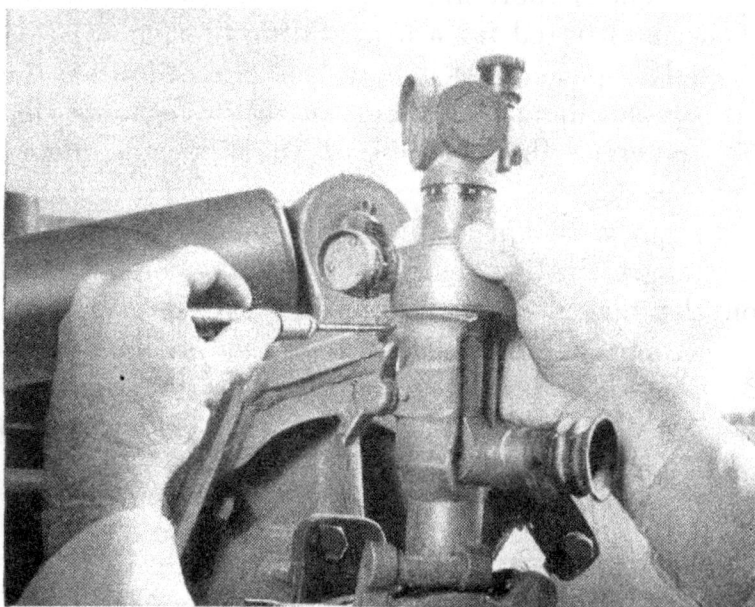

Figure 39. Tightening tangent screw on telescope mount.

b. Elevation adjustment. (1) Place the horizontal hair (optical center) of the reticle on the proper horizontal line of the testing target (distant aiming point) by turning the elevation knob of the telescope (fig. 40).

(2) If the elevation micrometer indexes do not coincide, loosen the clamping screws (fig. 41), and shift the movable index into coincidence with the fixed index (fig. 42). The elevation course indexes cannot be adjusted by using personnel.

(3) Tighten the locking screws and verify the line of sight and the matching of the indexes (fig. 43).

Figure 40. Placing horizontal hair of reticle on testing target by turning elevation knob of panoramic telescope.

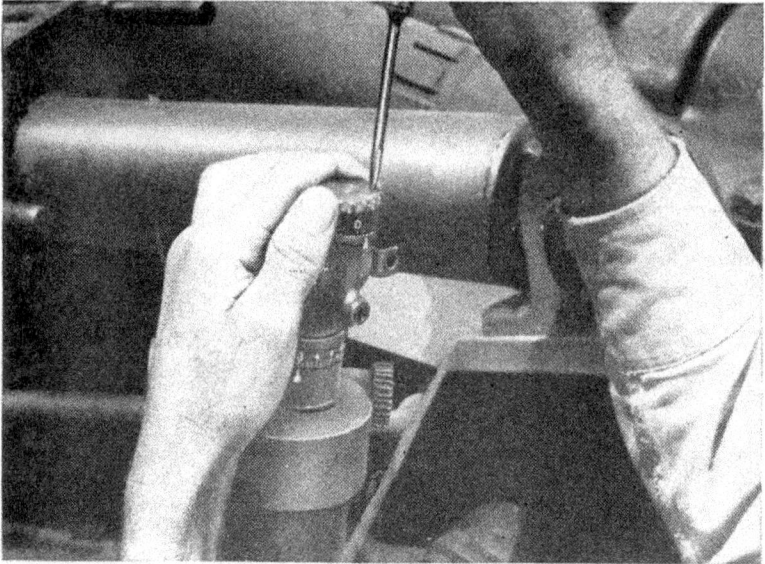

Figure 41. Loosening elevation knob clamping screws on panoramic telescope.

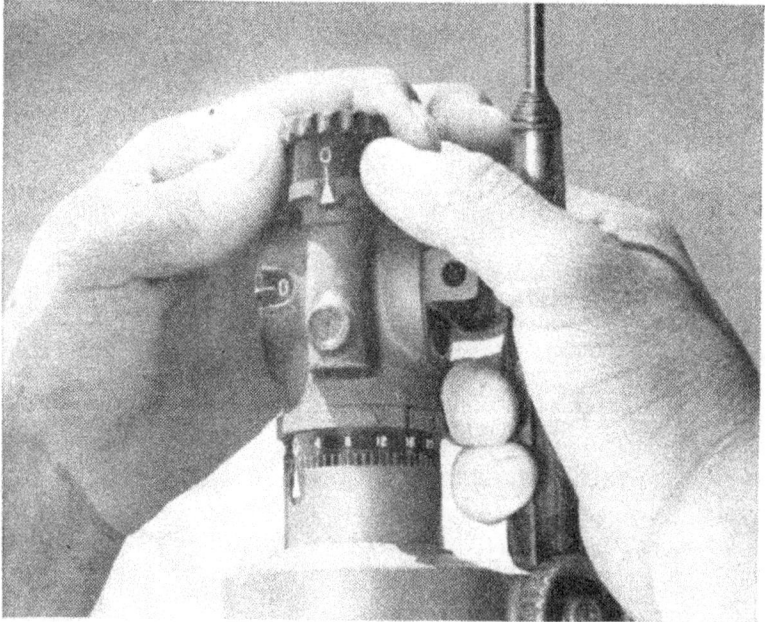

Figure 42. Shifting movable index into coincidence with fixed index on elevation knob of panoramic telescope.

Figure 43. Rotating head of panoramic telescope, indexes in coincidence.

c. Deflection adjustment. The vertical plane of the line of sighting can be adjusted parallel to the bore with the azimuth scales at zero by two methods. In either case, turn the azimuth worm knob of the panoramic telescope until the vertical hair (optical center) of the reticle coincides with the proper vertical line of the testing target or distant aiming point (fig. 44).

(1) *Adjustment by use of tangent screws.* (*a*) **If** the azimuth and azimuth micrometer scales do not read zero, turn the azimuth worm knob until the zero of the azimuth scale is in coincidence with its index.

(*b*) Loosen the three locking screws in the azimuth micrometer knob (fig. 45) and, while holding the

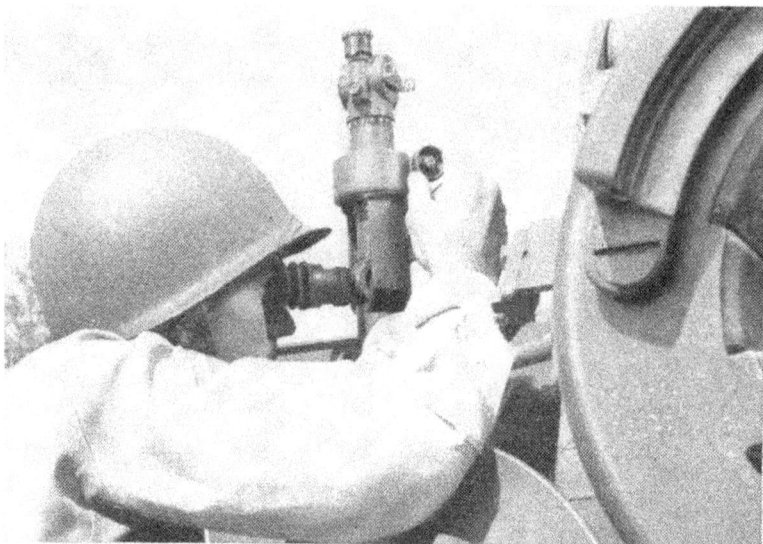

Figure 44. Placing vertical hair of reticle on testing target by turning azimuth worm knob.

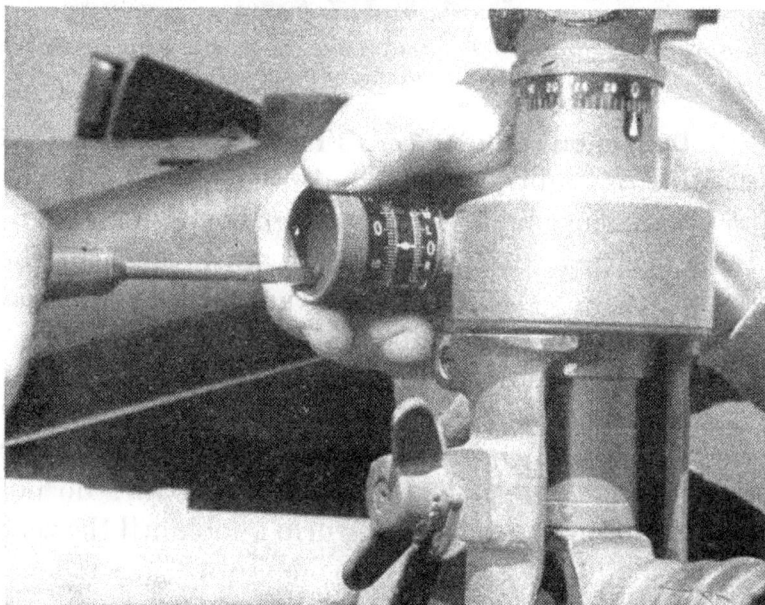

Figure 45. Loosening locking screws in end of azimuth worm shaft knob, preparatory to adjusting.

knob, shift the azimuth micrometer scale until the zero graduation is in coincidence with its index (fig. 46). Tighten the screws and verify the zero reading of the scale.

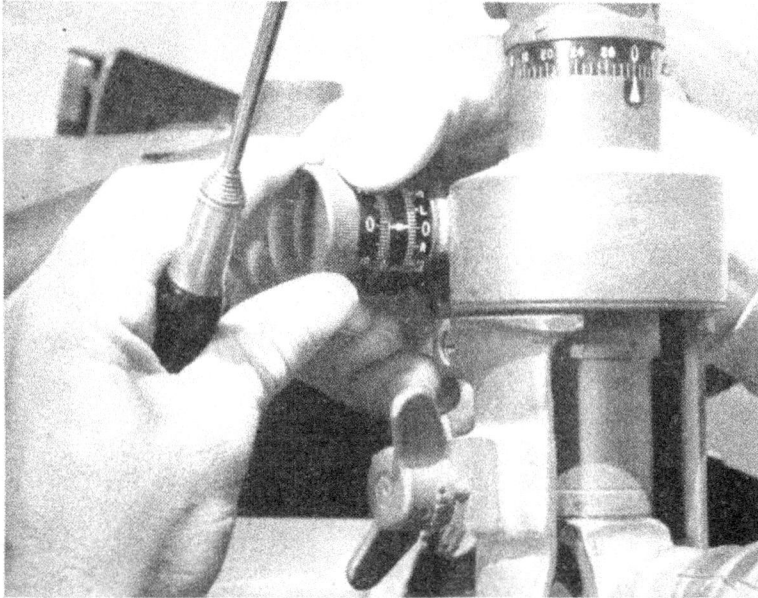

Figure 46. Shifting azimuth micrometer scale into coincidence with index.

(c) Loosen the tangent screw locking screws at the front of the socket and adjust the tangent screws (fig. 39) until the vertical hair of the reticle coincides with the proper vertical line on the testing target (distant aiming point), making sure that the panoramic telescope fits snugly against the tangent screws without binding. Tighten the locking screws and verify the zero readings and line of sight.

(2) *Adjustment by shifting the azimuth scale.* (a) If the azimuth and azimuth micrometer scales

do not read zero, adjust the azimuth micrometer scale as in **c**(1) (*b*) above.

(*b*) Loosen the locking screws on the azimuth scale (fig. 47) and shift the scale until the zero graduation is in coincidence with its index (fig. 48). Tighten the locking screws and verify the line of sight and the zero readings(fig. 49).

Note. The azimuth scale will not be shifted unless it is impossible to make the adjustment by use of the tangent screws.

Figure 47. Loosening locking screws which hold the azimuth scale.

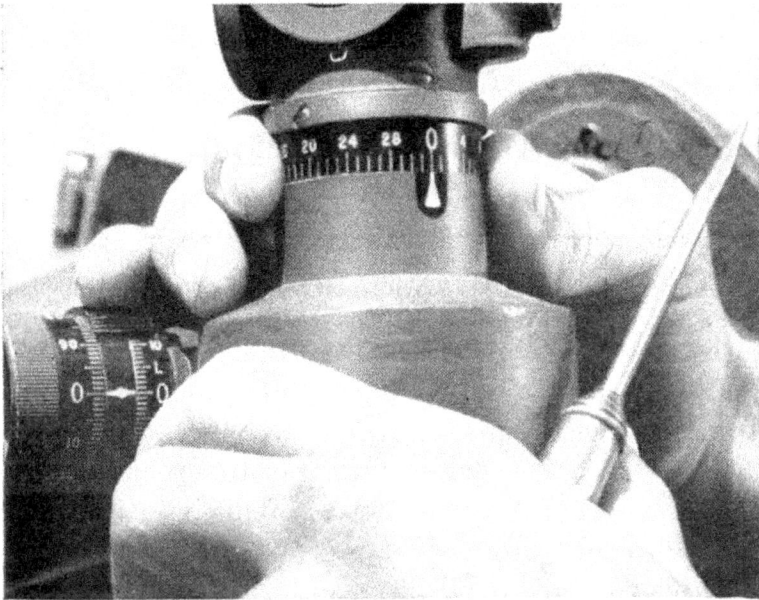

Figure 48. Shifting azimuth scale into coincidence with index.

Figure 49. Azimuth scales of panoramic telescope, showing scales in coincidence with indexes.

APPENDIX III

DECONTAMINATION AND DESTRUCTION OF MATÉRIEL

Section I. DECONTAMINATION OF MATÉRIEL

1. DECONTAMINATING PROCEDURES. a. For ammunition. Wipe off visible contamination from projectiles with rags. Apply DANC (decontamination agent M4), wipe with gasoline-soaked rag, then dry. If DANC is not available, scrub with soap and cool water. Wet mix or slurry (one pail of water to six shovelfuls of chloride of lime) can be used on contaminated ammunition containers, but it must not be allowed to penetrate to the ammunition itself.

b. For instruments. If exposed to corrosive gases, clean as soon as possible with alcohol (or gasoline, if no alcohol is available), and apply a thin coat of light machine oil. A rag dampened with DANC may also be used, followed by drying with a clean rag and application of machine oil. (DANC injures plastic or hard rubber surfaces.)

c. For weapons. Remove dirt, dust, grease, and oil. Do not apply wet mix but allow surfaces to aerate after soil and dirt have been removed. DANC can be used on all metal surfaces except the bore. Hot water, cleaning solvent, or repeated applications with gasoline-soaked swabs are also effective. If the emergency use of gasoline-soaked swabs is made, extreme care must be taken in order that the gasoline does not spread the contamination or that no gasoline in liquid or vapor form remains that may be ignited during future firing. After decontamination, weapons are dried and oiled.

d. For automotive equipment. Light contamination from spray can be decontaminated by aeration alone. For heavier contamination use DANC on interior or exterior surfaces which personnel are likely to touch. For larger area decontamination, wash vehicle with water and scrub painted surfaces with soap and water.

2. REFERENCES. For further information on decontamination, see FM 21–40 and TM 3–220.

Section II. DESTRUCTION OF MATÉRIEL

3. GENERAL. a. When capture by the enemy is imminent, matériel will be destroyed. *Such destruction is a command decision to be carried out only on authority delegated by the division or higher commander.* All howitzer batteries will prepare plans for destroying their matériel. Plans must be flexible in the time, equipment, and personnel required.

b. Training will not involve actual destruction of matériel, but will stress that—

(1) Matériel will be destroyed only when such action is necessary in the judgment of the military commander concerned. See **a** above.

(2) The sequence laid down for the method selected will be strictly followed to insure uniform destruction.

(3) The same essential parts will be destroyed on all weapons or vehicles to prevent the enemy from reconstructing a complete weapon or vehicle from several damaged ones.

c. Some methods require special tools and equipment not normally items of issue. Special issue of such items is a command decision. If blasting machines are not available, the generator of a standard field telephone can furnish current for tetryl electric blasting caps. With nonelectric blasting caps, at least five feet of Bickford safety fuze must be used to allow personnel firing to reach cover. See FM 5-25 for demolition charges.

d. Methods are given in the order of their effectiveness. Use Method No. 1 where possible, other methods in the priority shown.

4. OPTICAL AND FIRE-CONTROL EQUIPMENT.
Remove all detachable equipment before destroying the rest of the weapon. If evacuation is possible, carry detachable items and thoroughly smash nondetachable items. If evacuation is not possible, thoroughly smash all optical equipment and burn slide rules, firing tables, charts, etc.

5. TUBE, BREECH, RECOIL MECHANISM, AND CARRIAGE.
When simultaneous destruction of all these items is impossible, destruction of the tube, breech, and recoil mechanism has priority.

a. **Method No. 1.** (1) Start draining oil by inserting oil release in filling hole in recuperator cylinder rear head.

(2) Place armed (safety pin removed) antitank grenade, or armed (safety pin removed) antitank rocket in tube about 25 inches forward of forcing cone, ogive end toward breech. Place rocket or grenade in center of tube, using a wooden adapter, or wrap in waste to center it in tube.

(3) Load piece with unfuzed HE shell, and charge 7. (Do not use HE antitank shell.)

(4) Fire piece from fox hole at least 100 feet to rear, about 20° off the line of fire. Elapsed time: 2 to 3 minutes. Danger zone: 500 yards.

b. Method No. 2. (1) Ram an HE shell as if for firing. Insert 30 one-half pound blocks of TNT in chamber. Close breechblock as far as possible without damaging safety fuze, or carry safety fuze through primer vent and close breech.

(2) As a substitute for the shell, either plug muzzle tightly with earth for 18 inches, or if plugging is not possible, insert more TNT blocks in bore.

(3) Place one unfuzed HE shell (with eyebolt lifting plug removed) beneath recoil cylinder and above traversing rack. Cover fuze opening with one-half pound of TNT. Fifteen one-half pound blocks of TNT may be substituted for HE shell.

(4) Detonate all charges simultaneously, using detonating cord, tetryl nonelectric caps, and at least 5 feet of safety fuze. Electric detonation methods may be used if available. Danger zone: 500 yards.

c. Method No. 3. (1) Fire one howitzer at the others from position 500 yards distant. Use HE or HE antitank shell. Two or more hits on a vital spot should suffice.

(2) Destroy last howitzer and carriage by best means available.

(3) Enemy salvage of some parts is probable with this method.

d. Method No. 4. (1) With tube near zero elevation, insert eight unfuzed incendiary grenades in tube and close breech.

(2) Ignite another incendiary grenade equipped with a 15-second Bickford fuze, toss it in the muzzle, and elevate tube quickly to its maximum elevation.

(3) Destroy remaining parts by other means. Elapsed time: 3 to 5 minutes.

6. PNEUMATIC TIRES. Tires must always be destroyed, even when time is lacking to destroy the remainder of the weapon or vehicle.

a. Method No. 1. Ignite an incendiary grenade under each tire. If used in combination with carriage or vehicle destruction by explosives, incendiary fires must be well started before charges are detonated.

b. Method No. 2. Deflate tires. Damage with ax, pick, or fire from heavy machine gun. Pour gasoline on tires and ignite.

7. AMMUNITION. Adequate destruction of a full battery load takes 30 to 60 minutes. See TM 9–1900 for safety precautions.

a. Projectiles. Stack together *horizontally*, ogive ends pointing in same direction. Remove eyebolt lifting plug from center shell in top row of each stack, pack blasting cap (with safety fuze) in fuze cavity, and ignite fuze. Danger zone: 500 yards.

b. Propelling charges. Destroy by burning. Spread powder on ground, and ignite by powder train to permit personnel to take cover.

c. Fuzes and primers. Remove from containers, or open the containers, and stack in small piles. Saturate ground with gasoline or stack cans of gasoline around piles. Throw all available inflammable material on the piles and pour on gasoline. Ignite and take cover.

8. VEHICLES. See the technical manual pertaining to the vehicle in question.

INDEX

★ U. S. GOVERNMENT PRINTING OFFICE: 1 948—7515 53/254

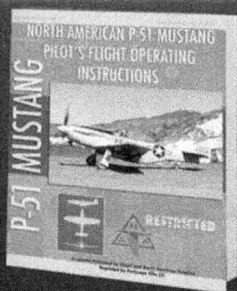

www.ingramcontent.com/pod-product-compliance
Lightning Source LLC
Chambersburg PA
CBHW052007090426
42741CB00008B/1592